ARMY

LO
B
Robert Taylor 85

DT A
Z865

The Smithsonian Book of FLIGHT

Walter J. Boyne

Smithsonian Books
Washington, D.C.

Orion Books
New York

Distributed to the trade by Orion Books, a division of Crown Publishers, 225 Park Avenue South, New York, NY 10003. Orion is a trademark of Crown Publishers, Inc.

Manufactured in the United States of America

Library of Congress Cataloging-in-Publication Data

Boyne, Walter J., 1929–
The Smithsonian book of flight.

Includes index.
1. Aeronautics—History. I. Title.
TL515.B665 1987 629.13′009 87-9699
ISBN 0-89599-020-2 (alk. paper)
ISBN 0-517-56614-1 (Orion Books : alk. paper)

First Edition
5 4 3 2 1

Page 1—Spiral of aircraft at the Smithsonian's National Air and Space Museum: Grumman F6F Hellcat, foreground; the Gulfhawk, *a Grumman G-22 aerobatics ship, at left; Boeing P-26A "Pea-Shooter" at upper right; Curtiss P-40E. Pages 2–3—Igor Sikorsky's* Ilya Muromets (1914), *the first four-engine bomber. Pages 4–5—Spitfire, left, and Hurricane in* Victory Salute *over postwar London. Pages 6–7—U.S. Navy McDonnell F-4 Phantom II jet after launch. Pages 10–11—Bumper crop of hot-air balloons rises over New Mexico. Pages 12–13—Piper Cub rests in its hangar after a perfect day's flying. At right; the U.S. Navy's Blue Angels in their Douglas A-4s. Curtiss SB2C during WWII, left. Overstuffed, unidentified aircraft, opposite.*

Contents

N1203R
Westgate

N7AX
VIRGO
LIBRA
SCORPIO
MATTHEWS
N1086R

N70682

Dedication to Paul Edward Garber NASM Historian Emeritus

At the Smithsonian, Paul Garber is synonymous with aviation. He is one of the few remaining members of the Early Birds of Aviation (who piloted an aircraft solo before World War I), and of the original U.S. Air Mail pioneers (who were in that service from May 15, 1918, until August 31, 1927), and he served in both World wars in flight-related capacities. Yet when we asked if we could dedicate this book to him, he wondered why we would want to do such a thing. He reminded us that there were many other air pioneers of greater renown than he.

We answered, "Paul, you knew a good many of the national and international figures personally, and since your life literally spans that of the modern age of flight, you are the Smithsonian's living memory of these exciting times. And more important than that, you have inspired so many people with your recollections and your enthusiasm for flight. Just this kind of continuity represents the Smithsonian at its best. It is such enthusiasm that keeps so many of our scholars and curators contributing to their fields for years after the usual retirement age."

Paul Edward Garber was born in August 1899, just three months after Wilbur Wright wrote to the Smithsonian Institution to express his intention to pursue the goal of powered, controlled flight with a manned machine. Four years later Wilbur and his brother Orville made history with their epic achievement at the windy seaside of Kitty Hawk, North Carolina. By 1909 the Wright brothers again demonstrated an advanced example of their Flyer for the military at Ft. Myer, Virginia. Paul was there and decided on the spot to seek a career in aviation. But we will learn more about that incident later in the book.

A member of the Smithsonian staff since 1920, Mr. Garber's greatest privilege was to bring the original Wright Flyer of 1903 back from a museum in England in 1948 and return it to the place of highest honor at home. In 1927 he realized the importance of the *Spirit of St. Louis* flight from New York to Paris and initiated the request that Charles A. Lindbergh agree that it be presented to the Smithsonian and added to its collections. As recently as 1979, Mr. Garber was associated with the accession of the first successful man-powered aircraft, Paul MacCready's *Gossamer Condor.* He assisted in the original planning and design of a building for the National Air and Space Museum, NASM, today the most popular museum in the world. The list of his accomplishments continues, not the least of which was the founding of the annual Smithsonian Kite Festival on the Mall in Washington, D.C. During the event's 20 years a legion of kite fanciers of all ages has participated. Though many of the youngsters have grown up, some bringing their own children to the event, they all treasure warm memories of windy days outdoors and hands-on experience with a real aircraft of their own.

"Don't forget that the enthusiasm for flight started at the Smithsonian long before me," Paul added. "Even James Smithson, our great benefactor from England, had an interest. For instance, in the 1780s he shared his investigations in mineralogy with a French scientist, Faujas de St. Fond, who was also a close friend of Jacques Charles, originator of the hydrogen-filled balloon. We can readily believe that Smithson through St. Fond would become interested in this development."

Mr. Garber's own involvement with flight at the Smithsonian began after his service in World War I and with the Air Mail. It continued until he was called to Navy duties the morning after the Japanese attack on Pearl Harbor. In the Atlantic theater he helped develop maneuverable target kites for ship-to-air gunnery practice and a kite-supported system for transferring special reports and equipment from ship to airplane to headquarters ashore. In addition, throughout the war, Commander Garber maintained the production of scale models of our own and enemy aircraft for use of pilots and gunners to recognize friend from foe. Back at the Smithsonian, he was the first curator of the newly established National Air Museum. Civil Service rules required his retirement at age 70; but he has continued in honorary active service as Historian Emeritus and Ramsey Fellow, the latter involving maintaining and improving of Navy materials in the museum. His latest project, not his autobiography but his "aerobiography," has just begun and promises to be the definitive account of the amazing progress of aviation during his lifetime and the Smithsonian's part in that adventure.

In a sense, by virtue of its long relationship with flight, the Smithsonian serves as the partner and elder brother of everybody whose life has been touched by aviation—and that includes virtually everybody who will read these words. Throughout his career at the Smithsonian, Paul Edward Garber has helped to focus the facts for millions of people who have been so greatly inspired by the descendants of that engine-powered Wright airplane of 1903 that grew up so quickly and sometimes so dangerously.

In fact, that ubiquitous word "modern" sometimes seems to have been coined especially to describe the marvelous period since the controlled and powered flights of the Wright brothers. And ever since they began the now universal advancement of flight, our old planet has never been the same. At the Smithsonian we thank our lucky stars that Paul Garber can guide us safely on our studies and explorations of humanity's highest adventure: to the skies and to the outer reaches of space. He has seen so very much!

Walter J. Boyne
and the Editors of
Smithsonian Books

Paul E. Garber, at right, and the Smithsonian Secretary pause briefly during discussions about the national aeronautical and space collections. They meet at the Secretary's offices in the Castle—the Institution's administrative center. Robert McCormick Adams, above, is the ninth Smithsonian Secretary: Mr. Garber joined the Institution during the tenure of the fourth, Secretary Charles D. Walcott.

AVIATION MEET
LOS ANGELES
JANUARY
10-20
1910
American & Foreign
Aviators
DAILY FLIGHTS

P R E F A C E :

The Once and Future AIRSHOW!

AIRSHOW! America's image of aviation has been focused through the airshow since the first gathering of almost surreal shapes in the sky at the fabled Los Angeles meet of 1910. The very word is redolent of the best of America. Airshow means bright blue skies, curving swirls of smoke from tiny biplanes doing impossible things, the blossoming of parachutes to dampen the heartbeat of a crowd tuned in to courage and danger. From the daring exhibitions of Lincoln Beachey and Barney Oldfield, down through the heavy metal unlimited races held in the golden years at Cleveland and now at Reno, airshows have conveyed the siren call of aviation, inspiring old and young with the thrill of flight. The magic has continued from decade to exciting decade.

The great air shows, no less than the auto races at Indianapolis or Daytona, have always been spiced with the glamor of hazard. In the old days, their underlying visceral appeal was of skill skirting tragedy, subtly reinforced by the prospect of catastrophe. But today, in the greatest airshow of them all, the Experimental Aircraft Association (EAA) Annual International Fly-In Convention and Sport Exhibition at Oshkosh, Wisconsin, all that has been substantially changed for the better. The performers are more skilled, the airplanes more beautiful, the danger present but deemphasized; the airshow has become a revolution in education, an inspirational guide to the true meaning of the freedom and adventure to be found in flying.

The very location is a help. Oshkosh has been the symbol of Americana for more than a century. Formerly known primarily for its overalls, b'Gosh, Oshkosh remains a movie-set town, rimmed by Lake Winnebago and bisected by blue rivers. People here have lived through the American cycle of agriculture, industrialization, decline and rebirth. All the while they have maintained their hometown's clean wide streets, huge frame houses on lots that call for carriage buildings and its frontier atmosphere of trusting amiability. In many ways, Oshkosh has become a rationalized Lake Wobegone, a place where Beaner's Shot and a Beer Tavern can comfortably coexist with granaries converted to gourmet restaurants. It is a resilient town, able to expand from its normal 50,000 population to accommodate the 750,000 good-natured people who swarm in annually during early August for the one hectic week of the Fly-In. Over the years, Paul Poberezny, the founder-president and guiding light of the Experimental Aircraft Association (EAA), has tuned the organization to match exactly the town's temperament, creating a symbiosis that might not have been possible in another setting. His instrument has been the matchless, ever-evolving Fly-In, an event of global renown.

Today the Oshkosh event is to former airshows as the

From the era of the legendary Los Angeles event, at left, to our own Space Age, airshow traditions are both celebrated and created at Oshkosh, the Wisconsin hometown of the Experimental Aircraft Association Fly-In each August. Above: Dale Crites of nearby Waukesha flies his re-created 1911 Curtiss Pusher above Oshkosh's Wittman Field.

Above, rolling to the airstrip, the EAA's own Ford Tri-motor delights Fly-In participants. "It flies like a pig," said the pilot of this clunky classic. Fifi, *the famed B-29 Superfortress of the Confederate Air Force of Harlingen, Texas, and other warbirds stand in the background. Top: Sue Parish of Kalamazoo, Michigan, polishes the family P-40N Curtiss Warhawk with the "Flying Tiger" on its side. Though pink was not "in" during World War II, this proud craft bears its original paint—its hue well mellowed.*

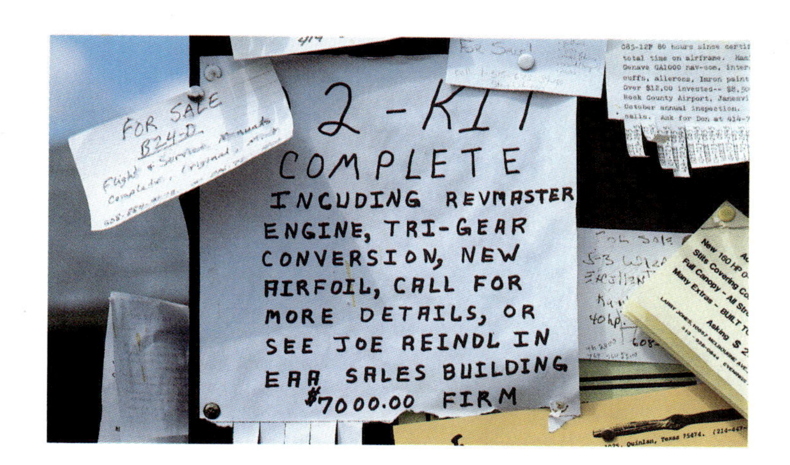

Left, bulletin boards help participants find others with kits, equipment or even entire airplanes to sell or swap. Thousands of Oshkosh's visitors camp out beside their aircraft for the entire Fly-In week.

Wagnerian opera cycle is to a street mime. It is an immensely satisfying multidimensional extravaganza, replete with Cellini-like craftsmanship, professional showmanship and a fervor of good will. This makes for a wonderfully warm visit absolutely unmarred by alcohol and drugs. Even litter on the ground is taboo at Oshkosh. Many of us believe that the annual Fly-In is the best possible example of camaraderie melded with morality.

In a manner that almost defies definition, the EAA has substituted statesmanship and education, draped in the magnificent panoply of superb craftsmanship and incredible flying, for the aura of real or synthetic danger of the airshows of the past. Today the Experimental Aircraft Association is the single most important factor in general aviation, not only for the enthusiasm its members generate, but also because of the ideas in design, construction and flight technique that the organization inspires. When the EAA began, it was not unusual for the experimental aircraft it espoused to consist of the wing of a Piper Cub shortened and mounted on a built-up fuselage. Today the EAA leads the aviation industry in terms of new materials, new engines and new methods; the gorgeous Starship to which the Beech Aircraft Corporation will commit billions of dollars to commercialize can trace its lineage directly back to the early aircraft with which Burt Rutan revolutionized the Fly-In.

Who could have guessed how far the EAA would come since the early Sixties. Yet it all happened by design, not chance. The EAA has become a leading factor in aviation education, with a very active and abiding emphasis on aircraft safety. The beginning was almost deceptively simple, a newsletter written and printed on a hand-run machine in Poberezny's basement; today it encompasses an enormous worldwide enterprise estimated to have brought $54,000,000 worth of business to Wisconsin. Through the years, Poberezny has demanded and achieved a degree of profes-

EAA founder and President Paul Poberezny, standing in front of the world's only flying P-64, takes time out from the day's events to visit with an EAA volunteer. Here, as throughout Fly-In days, good fellowship and Midwestern hospitality shine through. Overleaf: Right-side up or upside down, the wing-walking team of Earl and Paula Cherry thrills the crowd with high-flying stunts.

BUICK

ARL CHERRY
BUICK

L
483872
W
48543
U
483514
F

sionalism from a largely volunteer group; his pride has become their pride, and their pride is expressed during eight days of education-filled glamor.

The astonishing Oshkosh airshow presents every kind of aircraft, from lovingly crafted homebuilts to B-29s to the Concorde. One moment the sky may be filled with thundering formations of warbirds from many countries and conflicts, all spotless, all offering gleaming reconciliation rather than threat; the next moment by the gossamer wings of droning ultralights, turning in sparrow-size circles, recollecting the first dreams of flight by the human race. Then there are the antiques, all but legend—civil aircraft from half a century and more ago—refurbished jewels of flight. But the show's nexus, its *raison d'être*, is the proud passage of the homebuilt aircraft.

These planes arrive in every size and shape, from swift, fighterlike bullets, capable of cruising speeds of 300 miles an hour and fully equipped for all-weather flying, to glorious aerobatic biplanes, gleaming with 30 coats of hand-buffed lacquer. And just as shapes vary, so do construction techniques. Every permutation of construction possibility is represented, from wood-and-fabric to carbon fiber, all with one thing in common: absolutely perfect workmanship.

Among the acres of aircraft, one can sometime find a crowd gathered around a plane that is simply ordinary in

As thrilling as they are rare, four B-17 Flying Fortresses hold formation at Oshkosh, opposite. Top: Conventioneers watch the afternoon performances in front of one of many centers devoted to different interest groups of EAA. Famed test pilot Bob Hoover, above, babies his "mustard" P-51 Mustang. A favorite at airshows, Mr. Hoover performs precise maneuvers in this magnificent warbird.

Above, White Dwarf *draws its push from the leg muscles of Bryan Allen, who gained international fame for flying the pedal-powered* Gossamer Albatross *across the English Channel. Delightful Cricri "Crickets" taxi for takeoff, at right. EAA lifetime member Michel Colomban, a French aeronautical engineer, designed these easy-to-construct aerobatic midgets, which made their New World debut at the 1981 Oshkosh Fly-In.*

From out of the past, at top, two Boeing classics emerge. Owned by the Planes of Fame Museum in Chino, California, they are the P-12 biplane and the P-26 "Pea-Shooter." The last might be called America's first modern fighter plane: an all metal monoplane, it fought invading Japanese in the Philippines in 1941.

appearance. Its very plainness is so unusual at Oshkosh that people stop to stare, as if to relieve their eyes from the unvarying perfection of the other birds.

And this perfection leads to problems, for there are Grand Champions to be chosen in a wide variety of classes, and judging is a serious and difficult business at Oshkosh. Here the process has been raised to a standard of competence and excellence that exceeds even Europe's great automobile *concours d'elegance*. The underlying principle: the airplane must be flown to and at Oshkosh; no pampered trailered beauties these, with mirrors attached to show chrome underpinnings. These are airplanes, meant to be used, and their incomparable beauty and workmanship must remain subordinate to their utility.

This rationale of use and beauty is sophisticated; it makes the judging as egalitarian as the show itself, for it means that the simple outpouring of money will not win a prize. There is another factor, and that is the degree to which the entry has been built or restored by its owner-entrant. Factory restorations are not excluded—but they earn fewer points.

The judging process is done by carefully trained experts who meet daily to discuss the results of their efforts and to reconsider points they might have missed. The result is an elaborately fair, perfectly documented system that chooses winners on the basis of a well-developed bible of judging criteria and standards.

Some of the considerations are surprising; a super-fancy paint job may work against an aircraft that originally came from the factory in more mundane colors. The installation of prime leather seats in place of original factory fabric can hurt. On the other hand, a slight ding in the fuselage of an aircraft known for the resilience of its bungee-cord landing gear is not taken as a negative factor; it simply reinforces the required utility factor.

A few years ago, a wealthy sportsman brought in a rare biplane, immaculately restored. In the air and on the ground, it was a crowd pleaser, and popular sentiment had rendered it

Above, with his successful round-the-world airplane, Voyager, *Dick Rutan does some "hangar flying" in front of the familiar "Brown Arch" and the Oshkosh FAA Tower. A young enthusiast, right, takes in dozens of aerial highlights each afternoon. Italy's* Frecce Tricolori *Military Flight Team appeared for the first time in North America during EAA, OSHKOSH 1986. Colors represent those of the Italian flag.*

a grand championship by acclamation. But the crowd did not consider the inside of the cockpit as the judges did. There, instead of the black matte-finished dials of the original, chromed instruments of modern manufacture gleamed "better than new." The owner went into a state of mild cardiac arrest when the judges saw the instrument panel, shook their collective heads and took away the points that would have won a grand championship. The airplane gained an "honorable mention;" there was hell to pay—but integrity triumphed.

Back from the flight line where the judging and the airshows go on, beyond the flea markets and the hundreds of commercial exhibits, there are the forums, as many as 50 a day, conducted in open-air tents by the premier experts in the vastly varied fields. Here the true democracy and the true purpose of the EAA is defined. Here beats the heart of the EAA's education effort, for in the forum comes instruction on everything from "Proof Loading Your Airplane" to "Spin Resistance Criteria for Aircraft Certification." Here one can find such people as state governors, brain surgeons, flying

farmers and jet pilots listening raptly to the drawl of an airline mechanic who has built his own aircraft. In between, others teach how to weld, stitch fabric, navigate or select a kit to buy to build a real, flying airplane at home. The instructors are informal, the students demanding, the results dramatic.

Always most important, the EAA is sustaining an interest in aviation in America that would otherwise have disappeared in the face of declining aircraft production and rising costs. The organization has stepped in to deliver the message that aviation is fun and full of adventure. It has taken the frayed mantle of thrills from the airshows of the past and changed it instead to a proud banner for the future. In a single week, the great Fly-In recaptures the spirit of aviation that was born with the Wrights, blossomed with Lindbergh and came to shining fruition with the space program; moreover, the membership generates a vitality that promises that the future will be more than the past. The EAA holds that flight is attainable for all. Had Orville and Wilbur been able to chart the course they wanted for aviation, it would surely have been in the pattern of the EAA.

After an airshow performance, pilot Duane Cole and aircraft owner Woody Woods ride along the flight line in vintage style. They receive the applause of tens of thousands of Convention participants.

PART 1

BY TOM D. CROUCH

"They thought you were a fake, you see." That was the way pioneer exhibition flier Becky Havens remembered it. "There wasn't anybody there who believed an airplane would really fly. In fact, they'd give odds. But when you flew, oh my, they'd carry you off the field."

It was the same everywhere. The notion that a human being might move through the sky at will aboard a machine that was many times heavier than the colorless, odorless, apparently insubstantial fluid that supported it—well, it ran counter to common sense. It stretched credulity to the breaking point. "Flight was generally looked upon as an impossibility," Orville Wright recalled, "and scarcely anyone believed in it until he actually saw it with his own eyes."

People reacted in very different ways to their first sight of an airplane in flight. Orville's brother Wilbur noticed an "intensity of enjoyment," and a "sense of exhilaration" among those who came to see him fly. Then there was the fellow who was seen wandering away from the Wright trials at Ft. Myer, Virginia, in 1908 muttering, over and over again, "My God! My God!"

A Chicago clergyman attending his first air meet thought that he had "never . . . seen such a look of wonder in the faces of the multitude. From the grey-haired man to the child, everyone seemed to feel that it was a new day in their lives." A reporter covering the 1910 Dominguez Field meet in Los Angeles noticed

Airborne in the 1901 glider, Wilbur Wright learns to fly, a valuable prerequisite too often neglected by would-be inventors of flying machines.

From Dream to Reality

that same look on the faces of those who were watching an airplane take off for the first time,

"Thirty thousand eyes are on those rubber tired wheels, waiting for the miraculous moment—historical for him who has not witnessed it. Suddenly, something happens to those whirling wheels—they slacken their speed, yet the vehicle advances more rapidly. It is the moment of the miracle."

Miracle indeed. Even in an age that has come to regard human journeys to the moon and the robot exploration of the planets as commonplace, flight continues to inspire very much the same sense of awe, wonder, mystery and power that it did when the airplane was new. Aviation, that most hard-edged of technologies, has somehow retained a component of magic. We can explain the physical mechanisms and procedures of flight in the cold, hard language of the scientist and the engineer, but miracle and dream are still the words that come most readily to mind at the sight of an airplane tracing a contrail across the roof of the sky.

That is really not so surprising. The invention of the airplane began with our envy of the birds and involved nothing less than the realization of the oldest and most potent of human symbols. To fly was to achieve mastery and control over the environment, to taste ultimate freedom, to escape earthly restraint. From the beginning, we placed our gods in the sky and made flight, the one gift we had been denied, the universal attribute of divinity. "The natural function of the wing," as Plato noted, "is to soar upwards and carry that which is heavy up to the place where dwells the race of gods."

In the fertile river valleys stretching from Egypt east through Mesopotamia and India to China, the gods were firmly ensconced in the heavens when history began. The divine nature of the sky remained an important feature of more sophisticated religious traditions, as well. Taoist holy men were carried up to the next world by cranes. The Ascension of Christ is one of the central articles of Christian belief. Mohammed was borne through the six heavens by the winged mare Burãq, and travelled to the seventh aboard a flying carpet.

Angels and other messengers of the gods were invariably winged. Satan, the fallen angel, was shorn of his wings and cast down from the heavens to join the demons in the dark places under the earth. Ask any young child for the location of Heaven. They will invariably point up. Fighter pilot/poet John Gillispie Magee, who was to lose his life fighting on a new kind of battlefield four miles up in the sky, expressed an ancient and universal longing when he spoke of "slipping the surly bonds of earth" to enter "the high untrespassed sanctity of space," where he could "put out my hand and touch the face of God."

It is only natural that human beings would dream of traveling into the sky to commune with their gods. The high places of the earth—Sinai, Fujiyama, Ararat, Olympus, Denali, Chomolungma—were our first sacred places. In low-lying areas, we built elaborate towers to facilitate communication with the gods. How much more satisfactory it would be to construct wings on which to soar aloft, rather than having to shout up to the heavens from below.

But those who would imitate the gods had to be prepared to suffer their wrath. To fly was to tempt the fates, to trespass on the preserve of the Almighty. The story of Icarus and Daedalus, the classic myth of the air, illustrates the dangers, as well as the wonder, of flight. If God had meant us to fly, so the saying went, he would have given us wings.

Then, quite suddenly, a handful of men transformed the oldest dream into reality. Is it any wonder that so many found it difficult to believe? Human beings could fly. Not very far or fast or high, as yet, but fly nevertheless. The ultimate barrier had been overcome. Human flight, the very definition of the impossible, had been achieved.

The gods had not given us our wings, so we made them for ourselves. The psychological impact of the thing was stunning—overwhelming. With the heavens open to us, we would never be quite the same. If we could fly, was any goal beyond our reach?

Alive, aloft and probably in love, an unidentified Parisian couple celebrates the glory of flight from the gondola of their beautiful balloon. The banners bear the Latin phrase Sic itur ad astra, *"Thus one goes to the stars."*

MONTGOLFIER IN THE CLOUDS
O by gar! dis be de grande invention. — Dis will immortalize my King, my Country, and myself;
We will declare de War against our ennemi; we will make des English quake, by gar:
We will inspect their Camp, we will intercept their Fleet, and we will set fire to their Dock-yards:
And by gar, we will take de Gibraltar in de air balloon, and when we have
Conquered d'English, den we conquer d'other Countrie, and make them all Colonie
to de Grand Monarque.
CONSTRUCTING OF AIR BALLOONS FOR THE GRAND MONARQUE
FOURTH SKETCH
Published as the act directs March 2 1784 by S. Fores No. 3 Piccadilly
a Companion to this in a few days

False Starts and Triumphs

Although the Wright brothers were neither the first to try nor to fly, their solution to the ancient problem of aerial transport was unique. It contrasted sharply with all previous attempts, from the brave but foolhardy tower jumpers to the genius of Leonardo's luminescent designs to the serious but flawed approaches of Lilienthal and Langley.

Among those who knew and cared, the designers and builders of aircraft, the brothers were well appreciated. Some revisionists deprecated their accomplishments, most notably Gabriel Voisin and, later, the partisan advocates of Gustave Whitehead, Clément Ader and other claimants to the place of primacy. It was of course natural that those engaged in patent litigation—Glenn Curtiss, Augustus Herring and others—took exception to the Wright claims.

Almost 75 years passed before an accurate portrayal of the true genius of the Wright brothers reached the public. These two men from Ohio started out as bicycle builders, trained themselves as fine engineers, built and flew gliders of their own making and finally added power and control to their proved airframe. As saddened as they had been by the initial public rejection of their ideas and by the interminable patent fights, Orville and Wilbur would have been surprised and pleased at our almost universal appreciation of their scientific acumen, diligent planning and verging on miraculously swift progress from hobbyists to masters of the problem of flight. Having outlived Wilbur by 36 years, Orville Wright probably sensed their true place in history. He died in 1948, well into the Jet Age and at the threshold of the Age of Space. For his last flight, in 1944, he took the controls of the marvelous "Connie," the big, fast, new four-engine transatlantic passenger plane from Lockheed.

Our admiration of the Wrights is far more than a nostalgic look back at a time when America was relatively simple and unstressed. Instead, it is a sophisticated appreciation of an almost unparalleled concentration of effort from two young men who functioned intellectually almost as one and were within four short years able to achieve the success that had eluded so many others. Their vision became a reality, revolutionizing not only the way we move ourselves and our goods but also how we think about ourselves and our world.

Flight has often been described lyrically as being a part of our basic dreams, and literature abounds with allusions to flight either by super- or preternatural means. Thus empty eggshells were to be coated with the morning dew and wafted to heaven, or vacuums were to be induced in thin-walled copper shells, or geese were to be harnessed to haul the

Joseph Montgolfier, one of two brothers who invented the hot-air balloon in France in 1783, stated that he could conquer Gibraltar with an aerial army. The British lampooned him, at left. Below, Sir George Cayley's kite-like "boy carrier" probably lifted a lad in 1849. Overleaf: Potpourri of historic lighter-than-air craft, see page 285 for details.

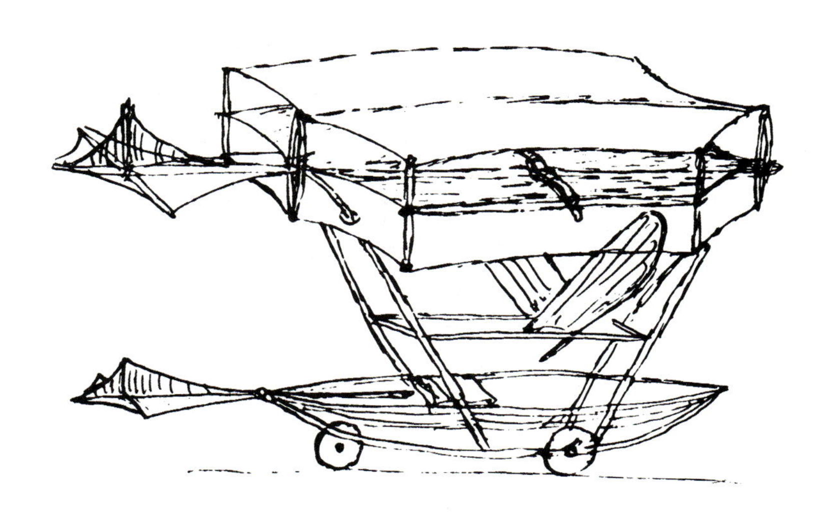

N.S.7
U.S.NAVY
NORGE
L3
HANSA
R 34

G
R101
G-FAAW
R101
G
AKRON
US NAVY
US NAVY
ZR-1

U.S. NAVY
GOODYEAR
NC-9A

Glider-pioneer Otto Lilienthal had a conical hill, right, built for him near Berlin. Standing atop it, hang glider at the ready, he could leap into a breeze from any quarter.

adventurer aloft. The curious thing is that manned flight actually had been technically possible for millennia, at least since the invention of the kite, probably more than 2,000 years ago in China or Southeast Asia. All of the materials necessary to make something resembling a modern hang glider have been available since ancient Egypt. An observant person, perhaps first working with models as later pioneers did, could have created a wood-and-fabric glider that could have soared safely from some prominence. Unpowered flight to be sure, but it could have occurred at almost any time. Perhaps it did in the very early years of the 18th century, for there are accounts of Father Laurenço de Gusmão, a Brazilian, who may have made a birdlike model that glided. One likes to think that if Father de Gusmão did not succeed, some other unknown person on a Quelin hillside made the great mental leap from kite to glider. Or perhaps, in Central America, a genius turned from crafting a calendar to building a quetzal-like flying machine. Why could not in the American West an Indian have arranged wood, bark and feathers in a way that sustained him in the air? Certainly there must have been many who tried and failed.

For that matter, if one thinks of the millions of people who have stared moodily into fires since the beginning of time, watching smoke and burning bits lift skyward, one can only wonder if somebody thought of the hot-air balloon. (It is said that Gusmão built one and demonstrated it to the Portuguese king in 1709.) Fabric and fire existed; all that was needed was profound insight, what patent experts call "original vision."

That view was gained by 1783 when, in the little town of Annonay near Lyon, two brothers were inspired to build the first balloon to harness the power of rising heat. Joseph and Étienne Montgolfier, scions of a still-famous papermaking family, were serious students of the art that foreshadowed the scientific approach of the Wrights. Their test program progressed from indoor experiments to a public demonstration of an aerostat—as such a device is called—on June 4, 1783. Fed with straw, fire heated the air to inflate the huge linen bag, 100 feet in circumference, and lined with paper. The big, open-air experiment suited France's venturesome intellectual atmosphere perfectly. The brothers were called upon to demonstrate their achievement in Paris before a sophisticated audience that included most of the members of the Academy of Sciences.

There they had a rival, J.A.C. Charles, a physicist who knew of Sir Henry Cavendish's experiments with hydrogen and who saw at once that this "inflammable air," weighing only a fourteenth as much as air, would be an even better lifter than heated air. Charles flew his invention, unmanned, on August 24, 1783, before a wildly cheering crowd that included Benjamin Franklin, who fully understood the importance of both invention and aphorism. (It was here that he made his oft-quoted remark about the putative usefulness of a newborn baby.)

Thus within a summer were born the two balloon technologies still in use today: the hot-air and the gas balloon. Events also signaled a soon-to-be-familiar phenomenon of flight: the

Lilienthal's remarkable series of more than 2,000 flights came to an end on August 9, 1896, with a crash that broke his back. Detailed drawings of 1895 Lilienthal-designed batwings appear below.

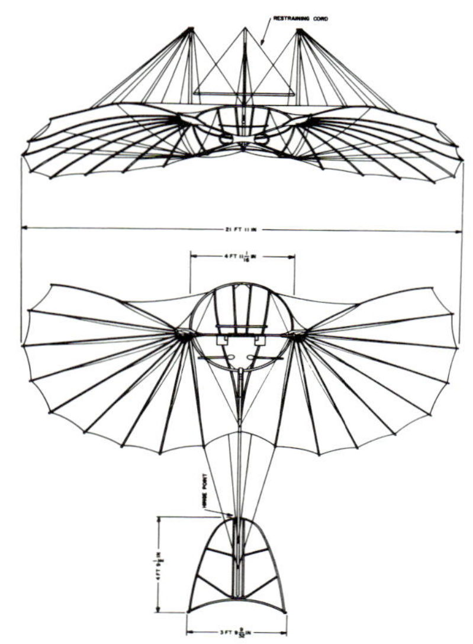

rivalry of competing technologies, each offering advantages and disadvantages.

The Montgolfiers had a command performance before Louis XVI and Marie-Antoinette at Versailles in September. More than a demonstration, it was an experiment to see if there were any unsuspected hazards in the upper air. (This was curious caution, as people had scaled mountains for centuries, but it typifies the scientific approach of the Montgolfiers.) A cage containing a duck, rooster and a sheep was attached to the balloon—it was before the days of the SPCA—and the balloon hoisted history's first passengers, for a flight of eight minutes and nearly two miles.

The first human flight—the royal proposal that two criminals be sent up was violently opposed on the grounds of the honor of France—was undertaken by J.F. Pilâtre de Rozier, on October 15, 1783. It was a tethered flight, only 80 feet up, but what a splendid first step into the sky! On November 21, 1783, de Rozier and the Marquis d'Arlandes made a free flight across Paris at a height of perhaps 300 feet, for a distance of a little more than five miles. (Today affluent vacationers travel across France in balloons, sipping champagne and stopping for waist-expanding dinners at châteaux along the route.)

However, it was the *charlieres*—as the hydrogen balloons invented by Professor Charles were called—that succeeded in gracing 19th-century skies and the public's imagination. Hot-air balloons were too dangerous. Two centuries later they would regain prominence, when the flames could be contained within a machined burner and the fuel was propane. Hydrogen balloons were also dangerous, but they offered far greater sustained lifting power.

Charles found that a relatively simple chemical process involving iron filings and sulfuric acid solved the problem of generating the hydrogen on a large scale. The first balloon comprised many of the elements found in a modern gas balloon: a car or basket fastened to a net enmeshing the balloon, a valve to release excess hydrogen and ballast to drop as needed to compensate for a loss of lift.

Despite the hazard now associated with hydrogen from the long series of 20th-century dirigible disasters, the early days of ballooning were relatively safe; one source asserts that there were only eight fatalities in the first 1,000 balloon ascents. The balloon was limited in capability, of course, for it was at the mercy of the wind, and there was no effective means of controlling its path above the ground. One went with and of the wind, choosing destinations dictated by the prevailing current. Jean Pierre Blanchard and Dr. John Jeffries took happy advantage of this on January 7, 1785, when they flew from Dover across the channel to Calais. No wind would be so politically laden until 124 years later, when Louis Blériot crossed *la manche* in the opposite direction in his spiffy Blériot XI monoplane.

Ballooning persisted as a sport for ever after, but in the 19th century the emphasis was on the profits gained by entrepreneurs providing short flights for passengers. In America, Jean Pierre Blanchard carried the first airmail letter—fittingly from President George Washington—on January 9, 1793. He had been preceded in the American air by Peter Carnes, who flew tethered flights in Bladensburg, Maryland, on June 24, 1784. Thomas Jefferson saw potential in the innovation ranging from ordinary cargo transport to meteorology, the discovery of the Poles and even such mundane and still-striven-for objectives as "raising great weights." Yet America was preoccupied with the great land rush that would carry it to the Pacific and beyond; with an assortment of wars, large and small; with despoiling the forest and dispossessing the native Americans; and with creating, in a fearsome rush, first a mammoth agricultural society and then a mammoth industrial society. Needless to say, there was not a central focus on ballooning.

Yet many intrepid experimenters were inspired by the dozen or so well-publicized flights made by Professor Charles Ferson Durant from 1830 to 1834. So great was the rush to emulate him that the press around Baltimore could comment in 1834 that "we are rather used to such things" and "balloon ascensions have become so common that we have ceased to notice them." The missing ingredient was tragedy, of course; the balloons did not blow up often enough to keep the press fascinated.

There were some great flights that received attention. Richard Clayton flew *The Star of the West* for 350 miles from Cincinnati to Monroe County, Virginia, taking care first to parachute a small dog from the basket on his ascent. The dog became so famous that he was jocularly considered popular enough to run for Vice President. And John Wise, veteran of 40 years of ballooning and 443 flights, made a record-breaking trip of 1,120 miles from St. Louis to Henderson, New York, in 1859.

All around the world, the balloon was periodically called upon for military duties. The first recorded use of balloons in warfare was by the French at the Battle of Fleurus in Belgium in 1794. In 1849, the Austrians attempted to bomb besieged Venice with timer-released bombs from balloons; uncontrollable, the balloons overshot, and dropped their explosives into Austrian lines instead. Thaddeus Lowe made great strides in the use of the balloon during the American Civil War, flying in some instances almost where the National Air and Space Museum now stands. The Museum exhibits a telegram from balloon-borne Lowe to President Abraham Lincoln.

The first significant military use of balloons came during the siege of Paris by the Prussians in 1870, when Félix Taurachon, popularly known as Nadar, organized a mail-and-passenger service on a very large scale. Some 67 flights were made, carrying out 155 passengers and nine tons of mail in balloons reputedly (but not really) made from the skirts of the ladies of Paris. Carrier pigeons became part of the outward-bound cargo, and for these a form of microfilm was devel-

Energized by steam engines generating a total of 360 horsepower, Sir Hiram Maxim's tripleton biplane foiled its metal restraining rails and almost flew in 1894. Interested only in scientific testing, Sir Hiram clipped his big bird's wings, in this version, and thereafter ran it low and slow—often for charity affairs.

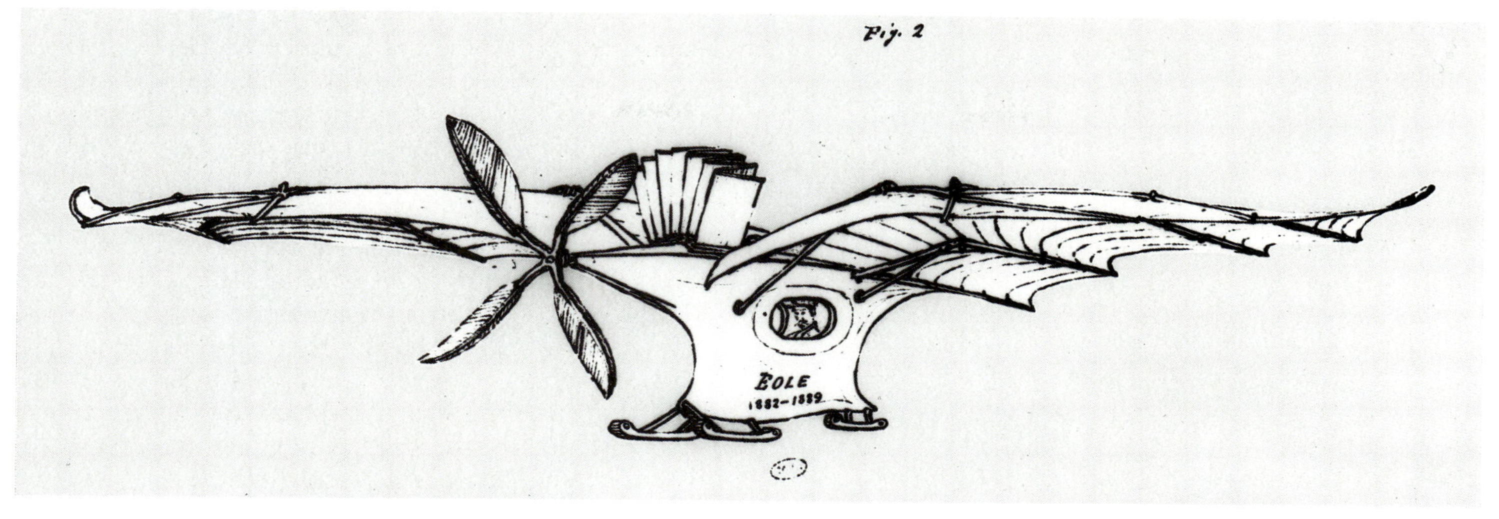

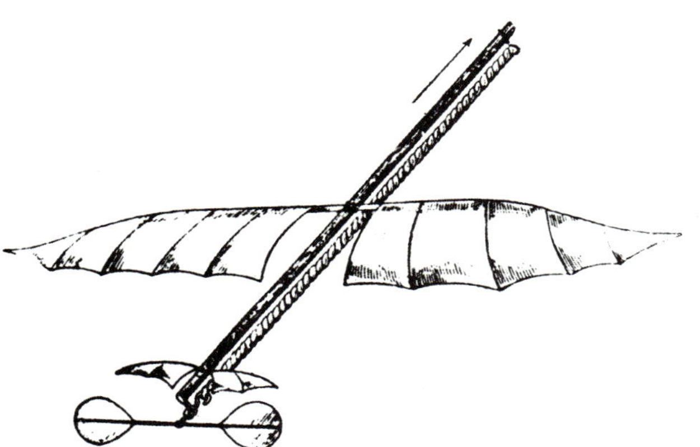

Clément Ader's Eole, *at top, may have made a brief hop under power in 1890. Twenty years earlier, Alphonse Pénaud's flying model, at left, covered a 131-foot circuit before the Société d'Aviation. This tiny* planophore *made early use of a rubber-band motor.*

oped for the return trip. Each piece of film had room for 2,500 letters of no more than 20 words each. Often fired at on their return journey by the besieging Prussians, the pigeons made the balloonists' improvisation, in effect, a two-way airmail service. A young German soldier named Otto Lilienthal may have been inspired by these flights.

But people were no more willing for their skyships than their oceangoing vessels to be ruled by the wind. Dozens of means of obtaining directional control of a lighter-than-air vehicle were tried. Some were self-defeating—a sail carried by a balloon is always in irons—and oars similarly had no effect. But as early as 1784 Jean-Baptiste-Marie Meusnier drew sketches of a balloon some 260 feet long, ellipsoidal rather than spherical in form. His proposed vehicle had three propellers to be turned manually, the first time that the screw had been suggested for propulsion of an aircraft.

In effect, balloonists had cast themselves adrift in the ocean of air in ungovernable cockleshells. It was not until Henri Giffard created the *dirigible* (meaning "steerable" in French) that the air became navigable at least in part. Giffard made a 17-mile flight from Paris to Trappes on September 24, 1852, averaging about five miles an hour. Run by a steam engine of three horsepower driving a propeller with three blades, the Giffard vessel clearly presaged the future, with its pointed, conically shaped body, suspended car for pilot and engine and triangular rudder at the rear. An anchor completed the nautical motif, suggesting one of the greatest problems encountered in lighter-than-air flight—that of ground handling, the mooring of the huge airships.

Giffard's flight set off an interest that corresponded to ballooning almost in direct proportion to the expense involved. It portended the giant dirigibles of Count Zeppelin that would dominate the lighter-than-air scene from 1900 to 1937, from Lake Constance to Lakehurst. Experimentation with dirigibles was more costly and more challenging than with balloons, and because of the requirement for motive power, far more hazardous.

The steam engines of Giffard and the electric engines of Captains Renard and Krebs in "La France" were too heavy, but the early internal-combustion engines were no more than time bombs for dirigibles. Early experimenters—Karl Wölfert and Jaegels Platz in Germany, the Brazilian Augusto

As third Secretary of the Smithsonian Institution, Samuel P. Langley, above, became a noted pioneer of flight. Though his man-carrying Aerodrome of 1903, above right, crashed twice, models, such as the one opposite, flew well. Right, his remarkably efficient gasoline motor of 52-horsepower, made by Stephen Balzer and Charles Manly, was the first radial engine fitted into an aircraft.

Severo in Paris and Baron Ottokar de Bradsky in Austria—all had disastrous accidents.

Alberto Santos-Dumont, that estimable if tiny Brazilian, brought the dirigible and internal-combustion engine into successful combination at the turn of the century by using his small airships as fiacres around Paris. And perhaps it was Santos-Dumont who inspired August Greth to make the first engine-powered airship flights in the United States. There had been a spate of human-powered airships, similar in concept to Bryan Allen's recent *White Dwarf.*

French by birth, Greth built the *California Eagle* to compete in the St. Louis Louisiana Purchase Exposition of 1904. The *California Eagle* engine, taken from an automobile, was too heavy and underpowered; while Greth did make the first airship flight in the United States on October 18, his vehicle was unsuccessful. Working with Greth, however, was Thomas Baldwin, who called on Glenn Curtiss for an engine for his own *California Arrow* airship. Baldwin stole the show in St. Louis, an event with incalculable consequence for both Baldwin and Curtiss.

Ah, the names and the associations of those early days! Captain Tom Baldwin brought Roy Knabenshue and Lincoln Beachey into the business; the three of them were dominant figures in the first decade of lighter-than-air work in the 20th century. Leaving their achievements and the later history of the dirigible to another chapter, however, we must return to the story of heavier-than-air flight.

Charting the relative progress of the two approaches is made easier when one realizes that dirigible pioneer Giffard was contemporary with Sir George Cayley (1773–1857), whose experiments earned him the title of the Father of Aeronautics. He was the first to attempt to explain the principles of heavier-than-air flight in mathematical terms.

He combined an insatiable curiosity with an intuitive yet systematic approach to problem solving. All science and nature was his domain: not only was he an engineer by repute and ability, he was also interested in everything from the streamlined shape of fish to engines fueled by gunpowder to helicopters. Cayley drifted in and out of aeronautical studies for more than 60 years, and if his attention span was short, his insight was deep. He was the first to understand that a vehicle of the proper shape, weight and surface area could, with a source of sustained power, achieve flight. And while he did not rule out the steam engine, during a quite remarkable flight of speculation he alluded to what came to be known as the internal combustion engine.

Cayley translated his vision into gliders and in his writings described the flights of several different vehicles. He included one that was supposed to have taken his unwilling coachman on a gliding flight in 1853. This reportedly caused the employee both fear and jurisdictional problems, because he had "been hired to drive, not to fly." Cayley's great insights were remarkable: the disposition of adequate, curved surfaces; the use of a pivoting, rear-mounted empennage or tail assembly; and the concept of dihedral for automatically inducing stability. Yet his greatest contribution was his influence on experimenters in his day and during the decades to follow. Through his varied experiments and his detailed reports of them, he legitimized the goal of flight and those who sought to achieve it.

This legacy established a climate that not only permitted William Samuel Henson to draw on Cayley's calculations and dare to dream of a large, profitable aircraft, but also allowed the British government to grant a patent for an "Aerial Steam Carriage" in 1843. Henson had a good notion of what an airplane must be, the shape it should take and the construction technique that should be employed. While some state that he did not have "original vision," his proposed aircraft nonetheless seems reasonable in appearance even today, with the single exception that it lacks the means of lateral control.

With Don Tate at left, Wilbur Wright observes the tethered flight of the 1902 glider. Once its vertical tail stabilizers were made to pivot, increased control was possible. Only a year away from success, the Wright brothers allowed that all they needed was to add an engine.

Henson's early collaborator, John Stringfellow, had shared with him the embarrassment of the failure of their Aerial Transit Company. It had proposed far more than it could have delivered (rapid transit to India and China, for openers). Stringfellow and Henson suffered the slings and arrows of an amused press. Henson gave up his experiments but Stringfellow persisted, ultimately flying a model with twin propellers and steampower that was similar to Henson's original idea. Hopping under controlled conditions, the model was history's first powered aircraft.

There followed a host of frustrated laborers in aviation's vineyard. Félix du Temple created first a model, then a full-size aircraft which, in 1874, made a hop carrying a sailor as a pilot. A similar feat was claimed for Alexander Fedorovich Mozhaiski in Russia in 1884. Both aircraft had used "assisted takeoffs," an inclined ramp, and neither was susceptible to control. But both left the ground with a man aboard, a giant step forward.

During 1890, the French engineer Clément Ader, aboard his batwinged, steam-powered *Eole,* possibly managed a takeoff, and without the aid of a catapult or ramp. If so, *Eole* may well have traveled the reported 165 feet at an altitude of a foot or less. Of "airplane size," with a wingspan of 45 feet and an empty weight of 498 pounds, the craft was energized by a 20 horsepower engine driving a propeller that resembled an eggbeater with splayed blades. There was an unbelievable oversight in a machine otherwise so elegant: it had no system of control. In design it combined unnecessary sophistication with very basic materials.

In Britain, the creator of the machine gun followed a similar but far more rigorous line of analysis, opting for a much larger scale. He could afford it! Sir Hiram Maxim had grown enormously rich on the earnings of his efficient killing machine. It not only made Victorian England even smugger ("we the Maxim gun have got, And they have not"), but it also utterly revolutionized warfare, raising by several degrees the hell that was World War I.

Maxim sought the keys to flight systematically, spending lavishly while proceeding cautiously. In perhaps an unfortunate metaphor for Maxim, he set his sights too low, seeking only to build a steam-powered vehicle that would leave the ground. This he did, with an enormous craft that had a wingspan of 104 feet. Two steam engines of his own design, lightweight and each producing a creditable 180 horsepower, energized his *Leviathan*. Weighing 3.5 tons, supported by a railway track, it succeeded in becoming airborne, though within the limits of the guardrails he had devised. Then Maxim inexplicably withdrew from the work that might conceivably have led to real flight before the 20th century. Such a success might have coupled the name of this expatriate American with a less lethal device or, even more likely, linked him to the first airborne machine-gun nest.

Almost all of the experiments to date had favored optimistic giganticism; the hopeful practitioners simply decided what sort of flying machine seemed reasonable to them

Dayton brain trust, below, Wilbur and his mustachioed brother Orville Wright developed the first manned aircraft capable of sustained and controlled flight. Testing, as with the motorless 1902 craft, took place on North Carolina's Outer Banks. This windy site, near both Kitty Hawk and Kill Devil Hills, held their wooden structure—both living quarters and workshop, bottom.

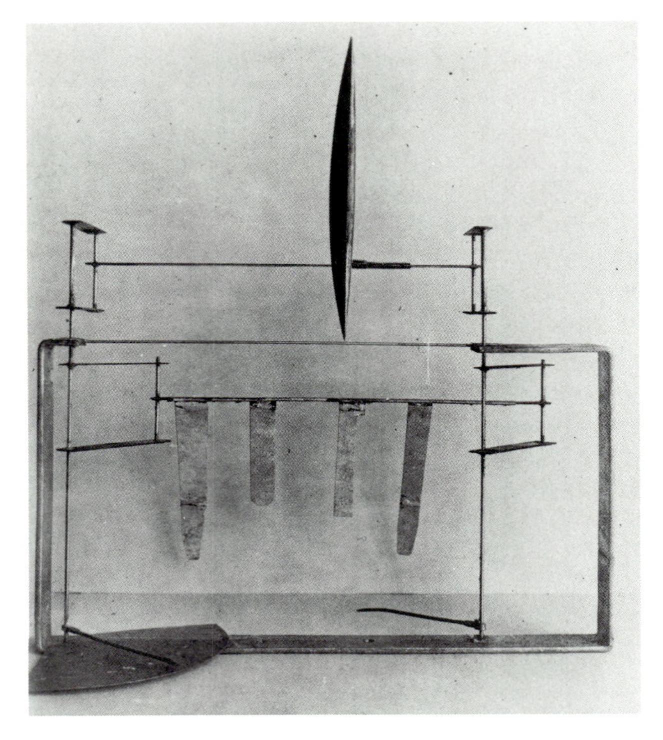

Exacting preliminary work at Dayton, including tests of wing-lift in the wind tunnel, came long before the assembly and testing of the Wright machines. The tunnel's balance, right, features a top bar upon which a scale-model wing with a curved cross section could be mounted vertically. Its lifting capacity was compared to those already calculated for the oblong plates fastened to the lowest bar. In this fashion, the Wrights compiled tables and diagrams of the lift of various miniature wings: the knowledge helped them design full-size wings. The entire wind tunnel appears opposite, the long device with the propeller.

and then built it as big and powerful as possible. Fortunately, all were denied success, for in no instance had they given sufficient thought either to the control of the craft or the skill of the pilot and, had any become airborne, disaster would surely have followed.

A new era began with the work of Otto Lilienthal in 1890. Lilienthal built his first glider—a hang glider in modern terms—and started a fruitful career that directly influenced the Wright brothers. Like most aviation pioneers, Lilienthal had been greatly interested in and affected by the flight of birds. This beguilement would lead him down the hopeless ornithopter path, but from his observations of flight in nature he drew inferences that gave him greater success than any previous practitioner.

There is a curious symmetry in the careers and relationships of the Lilienthal brothers, Otto and Gustav, and the Wrights. Each brother reinforced the other, and both teams made critical inferences about the flight of birds that their mechanical training enabled them to translate into practical air vehicles. This is a larger task than might be supposed, particularly when one considers the vast numbers of birdlike machines that were built with inherent structural flaws. These two duos were similarly careful experimenters, working from the known to the unknown on a very cautious basis, and extending themselves only after they felt their knowledge of the previous step was secure. (One wishes that more were known of the relationship of the brothers Montgolfier to see if the parallel still applies.)

Octave Chanute's engineering capability perhaps exceeded his judgment of human nature. He employed as his pilot Augustus M. Herring, one of aviation's gadflies. Many of Herring's contemporaries would regard gadfly as too kind a term but, as we shall see later, he must have possessed qualities of sufficient interest to attract the patronage—albeit always temporary—of a series of distinguished American aviation pioneers.

Chanute's prestige as an eminent engineer lent weight to his publications. Furthermore, his willingness to serve as a clearinghouse for ideas led to a transfer of leadership in aviation from Europe to the United States. Many epitaphs might have been chosen for a man with so distinguished a career as Octave Chanute, but the best of all might have been: *He was there when the Wrights needed him.*

Notable experimenters emerged during the latter part of the 19th century; Percy Pilcher, Alphonse Pénaud, Jean-Marie Le Bris, Professor J.J. Montgomery, Capt. Ferdinand Ferber and others. They were not part of the line of progression, though, that lurched to the Wrights.

The labors of the two brothers from Dayton have been well documented, and we now know of the extensive, systematic, rigorous scientific progress that they made in four incredible years of both inspiration and perspiration. Their achievements: the wind tunnel, the concept of three-axis control, their estimates of the necessary wing area, their establishment of accurate tables of lift of various airfoils, their realization that the propeller was a rotating wing, the creation of an adequate engine and the long practice in gliding flights to learn to fly. All these vital accomplishments have recently been emphasized in a number of excellent publications.

But how close were the other experimenters? Did Gustave Whitehead fly? Did Herring's fragile aircraft take off on a compressed-air engine? Was it only a launch failure that kept Samuel Langley from being first? Was Mozhaiski's flight a real one? Did Ader actually fly the *Avion III*? What of Richard Pearse in New Zealand?

It is unfortunate that a sense of advocacy divides historians

Living in the early days of the gasoline era, the Wrights ran both their wind tunnel and the belt-run, bicycle-shop machinery with primitive piston-powered motors. For the 1903 Flyer itself, they married bicycle technology to a box-kite glider and whirled its twin propellers with an internal-combustion engine of their own design and construction; the results amounted very much more than the sum of the parts.

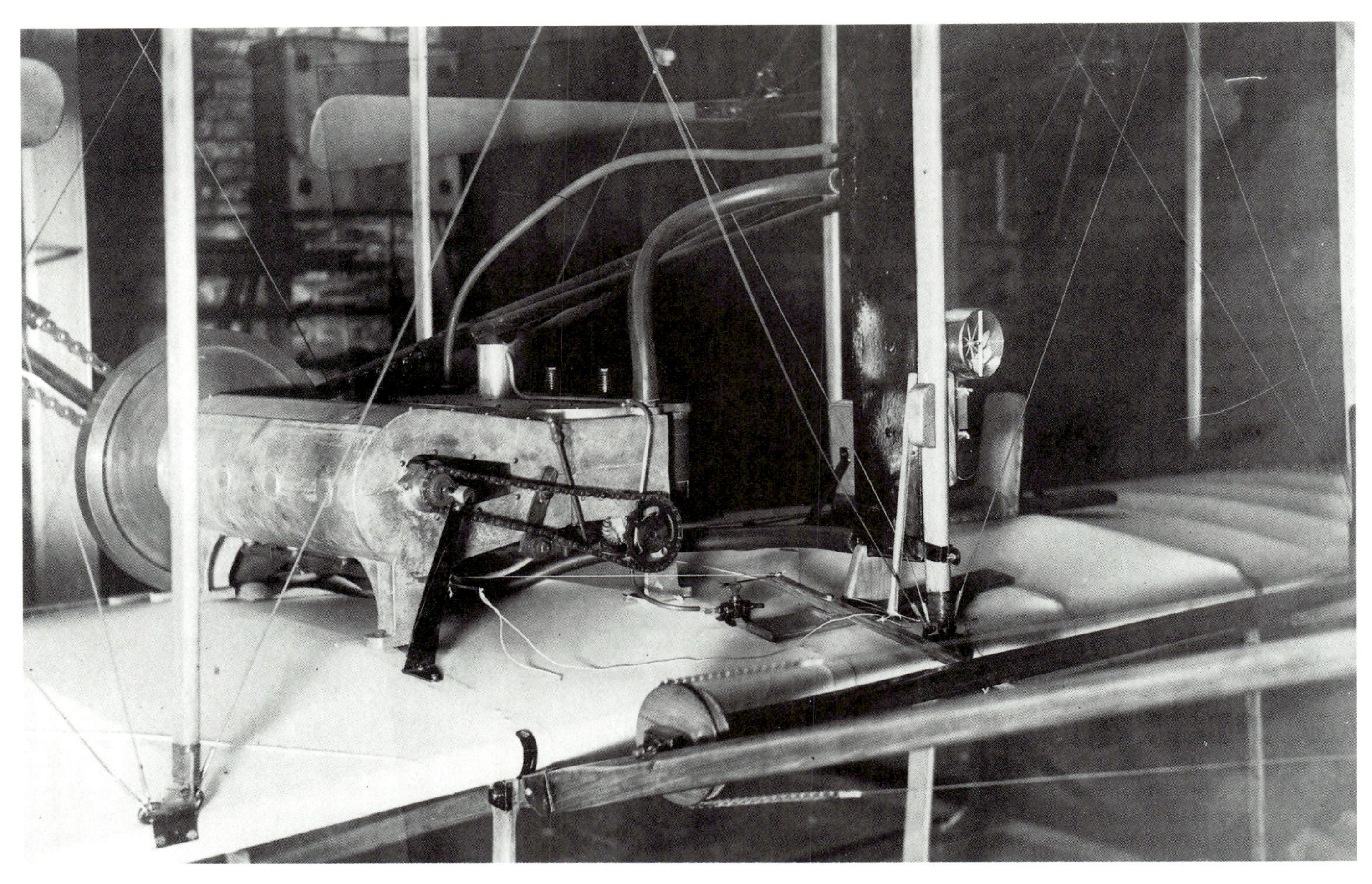

Historic photograph, at top, provides a close-up view of the 1903 aircraft's combination engine area and cockpit. The pilot lay beside the four-cylinder engine, its crankshaft connected to sprockets and chains to power the counter-rotating propellers. On the right-hand upright, a round anemometer measured air speed—it was loaned to the Wrights by Octave Chanute, a distinguished American civil engineer and air pioneer who visited the Kitty Hawk site, and is seen here, seated on the left. Chanute's portrait appears opposite.

into either-or positions on the value of the contributions of these pioneers. To say that everyone else in aviation was on the wrong track and that the Wrights were pursuing the one true faith would diminish the real value of their achievements. In fact, each of the pioneers, no matter what his lack of success, no matter if he scored an unrecorded triumph, were all important parts of the essential critical mass that resulted in the famous first flight at Kitty Hawk. This epochal event of December 17, 1903, had certain unassailable characteristics: a piloted, powered aircraft that took off without assistance from either a catapult or gravity and flew under control to a point as high as that from which it started. Furthermore, it unmistakably pointed the way to the explosion of flight technology that continues to this day.

Otto Lilienthal's machines were well made, composed almost entirely of materials that had been available for centuries, such as willow wands and waxed cotton. Lilienthal augmented gravity by always launching into the wind, often from a specially built conical hill near Berlin. Perhaps the first true pilot, a veteran of many flights, he recognized that he was engaged in a hazardous occupation. Though much more advanced than all others of the time, he had a conception of flight that remained imperfect because he depended upon the shifting of his weight for control. Rather than providing positive guidance, the pilot reacted to the aircraft's movements, compensating rather than initiating. Such a control system is fatally flawed, actually amplifying the effect of any misjudgment; that is, if a nose-up movement occurred, and he was not able to overcome it by shifting his weight forward, the force of gravity exacerbated the situation by throwing his weight back. Further, it put an upper limit on the size of aircraft he could build, because the ability of weight shifting to effect any control at all diminished rapidly as the size of the glider increased. (The modern hang glider came into being with weight shifting as the means of control, but today the pilot hangs in a harness below the aircraft in a vastly more efficient arrangement.)

More than 25 years of study passed before Lilienthal began his actual gliding experiments in 1891; then in five short years he made more than 2,000 flights before he was fatally injured in a fall to the ground in 1896. By then he had designed seven different types (five monoplanes and two biplanes) and had advanced aviation further than all previous practitioners. But even more than his intellectual contributions, Lilienthal provided a shining example of the engineer-pilot, the man who proved his theories in practice, and before cameras. Images of bird-winged gliders aloft, with his legs cocked beneath, were seen around the world and inspired others to follow, including Octave Chanute. Yet it may be that Chanute's most valuable contribution was the seed he planted in the minds of Orville and Wilbur Wright.

An important forerunner of the Wrights, Octave Chanute possessed a formidable engineering background gained in the forefront of railroad technology. He used his considerable powers to infuse developments in Europe with his own concepts of control.

Chanute felt that "automatic stability" should be sought and that movable surfaces were a better approach to control than shifting of the aviator's weight. He took the baton from Lilienthal's fallen hand in 1896 by beginning a series of only moderately successful gliding experiments. Despite his intuitive preference for movable controls, for practical reasons he persisted with weight shifting as his control mechanism.

In addressing the claimant pioneers, one sometimes becomes trapped in emotion and sentiment. A sincere and enthusiastic group asserts that Gustave Whitehead made the first flights in the United States in 1901. Many people do not feel that there is sufficient evidence for this assumption, but neither point of view really matters. The measure of the man is not in public acclaim but in the amount of dedication and effort he placed in his work.

Whether or not Whitehead flew, he made a stimulating—and, quite inadvertently, a continuing—contribution to aviation. Similarly, if the Wright brothers had not flown, the human value of their role as experimenters would not be diminished. What must be applauded is each contributor's commitment to an ideal. And in these later years a mature assessment must be made not merely based on who was first, but rather on the total effect. In the next chapter we shall

---

Overleaf: Aviation's premier moment, December 17, 1903; Wilbur Wright makes the first successful takeoff in the Flyer. This 100-foot leap has lead humanity toward the stars. The day's final flight of 59 seconds traversed a distance of 825 feet.

Gleaming at the National Air and Space Museum, the original Wright Flyer of 1903 occupies the place of highest honor in its home country, and in the annals of United States aeronautical history. Opposite, in 1911, Orville Wright returned to Kitty Hawk for the last time—this time to establish a world's gliding record of nine minutes and 45 seconds. He foresaw the day when highly engineered aircraft without engines could emulate the condor, harnessing the wind and sailing for hours effortlessly.

attempt to form a true picture of the accumulated strength of these pioneering efforts.

Yet the thirst to be first ran deep, and bitter was the failure to slake it. Perhaps the saddest of all the "almost first" pioneers was Samuel Pierpont Langley, third Secretary of the Smithsonian Institution. A distinguished scientist, in 1896—the year of Lilienthal's death—he launched large steam-powered models on many successful flights.

Like Maxim, Langley originally set his sights too low; having flown models successfully, he assumed he had proved the possibility of manned flight and turned to studies of a different nature. In 1898, though, his patriotism was stirred—and perhaps his pocketbook lured—by the offer of financial support from the government for the creation of a military aeroplane. He returned to aircraft construction and by 1901 had built a gasoline-powered version of his tandem-winged model. Its flight marked the first time that such a power plant had propelled an aircraft.

What happened next now seems implausible, given Langley's scientific background. Somehow this well-financed astronomer made a series of fantastic assumptions. The first was in method; he simply scaled up his models to create the full-size aircraft, not understanding that strength of materials and construction methods do not scale up in direct proportion to size any more than speeds increase in direct proportion to power. The second was in insight; he assumed that the Aerodrome, as he called his invention, would be automatically stable and that the pilot would simply steer it about like some winged rowboat that could ply the ocean of air. Third, he failed to provide an undercarriage, electing to launch the vehicle from a houseboat just as he had done his models. Thus his brave pilot, Charles Manly (who also built a wonderful water-cooled radial engine for him) could expect either to be submerged or abraded, depending upon whether he landed wet or dry. Fourth, his launcher was not perfect, as he knew from his experiments with models. Finally, Langley did not appreciate perhaps the most obvious need of all: the pilot should know how to fly. Manly had no opportunity to familiarize himself with the Aerodrome even to the limited extent of taxiing before being flung off the houseboat like a giant clay pigeon.

Langley created a superbly elegant machine, brilliantly constructed and based on successful models; but it had no chance whatever to fly. It was doomed to structural failure from the start, and Manly's survival of two crashes is as much a credit to him as is his decision to try again after the first.

In a way it was Langley who did not survive the second crash, which occurred under the full glare of publicity only eight days before the Wrights' successful effort. He died in February 1906, a broken man, hurt not so much by his failure to do what the Wright brothers did but by the deluge of ridicule inflicted on him by the press.

The Wrights had absorbed Lilienthal, exceeded Chanute and outengineered Langley. They flew in 1903, and the world ignored them—but only for a while. Then sadly, inexplicably, in the midst of success, they ignored the world.

America's Forfeited Legacy

In the Americans' early disbelief and disinterest in the accomplished fact of flight lay the seeds of misfortune. Aviation in the United States was relegated to the status of sideshow exhibition. Furthermore, the Wrights' retiring, American-Gothic personalities did little to promote responsible air-mindedness.

Yet the Wrights had certain qualities of mind and will that enabled them to make their historic breakthrough, to gain the original vision of flight as we know it and to triumph in spite of popular disinterest. And when people took to watching air stunts for entertainment, the Wright brothers helped organize a troupe and took a share of the instant wealth.

As it happens, typhoid fever serves as an odd and ghastly bracket to the achievements of the Wright brothers. That dreadful disease is rare in most of the Western world today, but not in 1896 when Orville was wracked with it. That was when he first became interested in the feats of Otto Lilienthal. In addition, his illness seemed to erase the brothers' difference in age, drawing them into a relationship that squared rather than doubled their combined intellectual power. When Wilbur died of typhoid in 1912, the fever broke up one of the closest working partnerships in history.

They began their work in earnest in May 1899 with a letter to the Smithsonian Institution. Orville's fever could not have imparted a vision of the full complexity of the problem to be addressed; for if it had, they might never have begun. Yet within only four years, using an intuitive scientific approach that achieved swift progress with spartan economy, they solved all the problems they encountered. Rarely has any team covered so much ground in so little time with such complete success.

They did it at a time when the very idea of manned flight was decried by science and by common sense, so much so that the actual event on December 17, 1903, passed almost unnoticed. People simply didn't pay attention to it. How many times did a solid citizen have to be stung with the announcement of an aerial locomotive, skyborne wagon or flying ship? Had not the most famous practitioner of all, Secretary Samuel P. Langley of the Smithsonian Institution, just demonstrated conclusively that heavier-than-air manned flight under power and control was an impossibility at least for the next century?

Yet succeed the Wrights did, and though aviation truly began with them it would soon take on a life of its own as serious development moved to Europe. The decade from 1903 to 1913 may be divided into two five-year periods. In the first, the Wrights made their aircraft practical, and were in fact ten years ahead of the world. By 1908, the watershed year, the world was forced to take notice. Orville made history's first flight of an hour's duration at Ft. Myer, Virginia, and seemed just about to secure the military contract that had hitherto eluded the brothers.

The Wrights had promised to deliver an airplane that

Only after Wilbur Wright's European Triumph of 1908 did Britain and the Continent put their aerial act together. Soon, though, the Old World quickly outdistanced the New, hints of which appear in a 1909 poster for a flying meet at Reims, opposite. With the American legacy lost, U.S. manufacturers followed overseas design, as with the Thomas-Morse S-4, above. An 80-horsepower French Le Rhône engine powered the pretty biplane.

The Wright Military "Flyer" at Ft. Myer, 1909

Paul E. Garber

When I was ten years old, back in 1909, my father brought me along on a business trip to Washington. He was an art dealer looking for a place to establish an art gallery. One evening after dinner we were reading the newspaper. On the front page was a picture of an airplane. The caption described it flying at Ft. Myer. "Oh, father," I said, "may I go see that?" From the hotel clerk we learned that Ft. Myer was across the Potomac in Virginia, on a local train line.

The next morning father took me with him to see some art dealers, but after lunch gave me carfare, and I was off to Ft. Myer. There I saw a huge drill field flanked by the Arlington cemetery wall along one side, many army buildings and stables on the far side, with residences and a hospital at one end. As the sound of the departing train diminished, I could hear a sound to my left. There in the distance I saw the airplane in flight making a turn and coming toward me. I was far more than surprised. I had seen the picture of it the evening before. I was elated and astounded! My Uncle Ed had taught me to make and fly kites, and our latest project had been a box kite. The approaching shape was somewhat similar but far more impressive and immensely larger. There were two men in it sitting on the front-center of the lower wing. Beside one was an engine and behind the man were two rotating swirls. As it turned to the left and headed toward the far corner of the field, I ran after it, soon getting into a large crowd and pushing toward the front edge, where some soldiers were standing with their guns horizontally down at arms' length as a barrier to hold the people behind a line. I wiggled to that front edge where some man put his hands on my shoulders,

letting me know that was as far as I could go. The airplane had flown past the far end of the drill field, where it had turned left approaching the cemetery wall. Then it turned again, repeating the course that I had first seen it fly. Its next turn was toward the crowd where I was standing. Turning overhead it again flew to the far end of the field. After several more circuits, it angled down and landed.

My shoulder-patting friend now showed a card to the nearest soldier, who recognized it as a pass and permitted him (and me, to my eternal gratitude) to walk over to the airplane. I was cautioned not to touch anything, and I obeyed with my hands clasped behind me. My sense of sight was amplified by my sense of smell as I stood near the engine and inhaled the fragrance of hot oil. I also imbibed that invisible but very potent sense that emanates from seventh heaven. I walked around the airplane absorbing all that I could.

Eventually some soldiers raised one wing-end of the airplane while another one guided a large wheel and frame under the wing near its center; then with the other wing-end raised, another wheel was placed on the other side of center. Carefully the crew rolled the airplane to the hangar at the far end of the field.

With my mind floating on air, I headed back to the city. My father was perturbed because I had been gone so long, and he was hungry.

He asked, "Well, how was it?" I searched for words as he repeated the question. I couldn't tell him. I was slowly shaking my head. Then I reached into my jacket pocket, took out the news article of the airplane, pointed at the picture and finally stammered out, "THAT's where I was!"

would be able to carry two people with a combined weight of 350 pounds, with enough fuel to fly for 125 miles, at a top speed of at least 40 miles per hour in still air. The craft was also supposed to have at least an hour in the air and to be able to land without causing damage that would preclude another immediate flight. For this they were to receive $25,000, with a ten percent incentive clause for each full mile an hour over the 40—and a matching penalty for each mile less.

All of their hopes were dashed to the ground with the airplane when—according to Orville's postflight analysis—a guy wire broke, damaging a propeller and causing a control problem. Lt. Thomas Selfridge of the Army perished in that September 17 crash, the first of many martyrs to powered flight. By July 1909, Orville was flying again and had developed an aircraft to meet all of the Army's requirements.

Across the ocean, Wilbur turned a scornful and doubting host of Europeans into flight groupies by attaining record distances and durations—and with astounding skill. But after 1908, suspicion, lawsuits and bitter rivalries would grind the triumphant achievements of the Wright brothers into the European dust.

The Wrights are now rightfully given their place in history as the inventors of the aircraft. Too often, though, their contribution to the utter abdication of American aerial supremacy is overlooked. The Wrights' belief that no one could catch up with their invention and their preoccupation with defending their patents hurt not only themselves, but all American pioneer aviators, particularly Glenn Curtiss.

It was perhaps human, after all their rebuffs, and after the scorn that was heaped upon them, that the Wright brothers should have acted as they did. They were in fact five and perhaps ten years ahead of the European experimenters in 1903; but while time unfolds in unvarying astronomical terms, progress advances by fits and starts. They lost nothing but perhaps sales in the close-to-the-chest policy they pursued from 1903 to 1908. Once they had set Europe afire in 1908, however, there was no way to check progress elsewhere. The Wrights forced Curtiss in the United States and a host of adventurers in Europe to cast wildly about for ways to break their lock on patents; meanwhile, their competitors made enormous progress. On the other hand, the Wrights were content to improve their aircraft gradually, without radical departure from their successful 1905 machine.

The Wrights worked through 1913 to develop their machines. The improvements they made were marginal, however, barely ahead of the pack, *just sufficient* to stay up front. As the Kitty Hawk aircraft had incorporated just enough wing area, just enough power, just enough control to do the job, so were the succeeding Wright aircraft *just enough* to appear to be competitive.

Theirs was economical engineering, but shortsighted in business terms. To illustrate, the Wrights had dramatically improved their engine in 1906; during the next few years, they simply tweaked it into a reliable power plant that in-

Talented motor maker, Glenn Curtiss, crept into flying, as evidenced on these pages with his early engine and control test vehicles.

creased in power from 25 to 30 horsepower. By 1913, to meet the need for increased power to carry experimental radios or extra fuel for extended flight, the company introduced a six-cylinder model that developed 60 horsepower. It retained the twin-pusher, chain-driven propellers, with their clattering and screeching, as well as the wing-warping control system that had served so well, but was now passé.

The reason for the Wrights' obsolescence—and for European ascendancy by World War I—lies in a quest for patent protection. Instead of competing, the Wrights expressed annoyance at the persistent efforts of other experimenters to subvert their primacy by the introduction of such devices as ailerons, controls and wingtip extenders.

The Wrights fought a long war and won many victories, all Pyrrhic. Over the years the courts in both America and Europe sustained their patents, but clever lawyers worked out arrangements by which Curtiss and others could continue their development work while contesting in court, or filing appeals. World War I mercifully put an end to the lawsuits when the U.S. government made it a patriotic necessity (and, incidentally, threatened to give all business to others) for the Wright and Curtiss companies to enter into a cross-licensing agreement. The agreement brought millions to both parties, but never eased the acrimony. Glenn Curtiss did not emerge from the fray unscarred.

Curtiss started out building bicycles, as did the Wrights, then became a record-setting motorcycle racer and, most notably of all, a creator of powerful, lightweight engines. As early as 1906 he had offered some for sale to the Wrights. He drew closer to a personal involvement with aviation when Tom Baldwin asked him to build an engine for an airship, and during 1907 he was most fortunately brought into an early-day "think tank," the Aerial Experiment Association (AEA).

Founded by Dr. Alexander Graham Bell, the Canada-based AEA was fully confident of making history, largely because AGB (Dr. Bell liked initials and abbreviations) had done so well with his invention of the telephone. He had also followed Langley's experiments and felt that a different approach was required, one involving large and complex tetrahedral kites. Dr. Bell and Curtiss were joined at the Baddeck, Nova Scotia, laboratory by other significant names, including F. W. "Casey" Baldwin, J. A. Douglas McCurdy and that same Lt. Thomas Selfridge who was to be killed at Ft. Myer in 1908 with the crash of a Wright machine.

It was a distinguished, versatile group to which Curtiss brought more than a mechanic's knowledge of engines. He had competed in both races and business, was familiar with trial-and-error experimentation, and valued not only the ideas of others, but also their ability to contribute to the success of a program. It is here that he perhaps differed most significantly from the Wrights. He also took much from the Aerial Experiment Association, not least of which was experienced inventor Bell's injunction to photograph and document all that he did in case of later patent fights.

With engine and engineering experience honed by motorcycle manufacture, Glenn Curtiss possessed just the know-how required by Alexander Graham Bell to make his own early bid for flight; see next page.

An early flight seeker, Alexander Graham Bell financed Canada's Aerial Experiment Association—from left to right: Frederick W. "Casey" Baldwin; Lt. Thomas Selfridge, the first person to die in the crash of a powered aircraft; Glenn Curtiss; Bell himself and John A.D. McCurdy, the first Canadian to fly. Augustus Post, who is not a member, appears at far right.

With June Bug, *below, Curtiss flew a kilometer course nonstop in 1908, winning a prize offered by* Scientific American *magazine. The publicity thrust Curtiss into the forefront of aviation—at least in the eyes of the public. While his plane was flyable, it was hardly controllable, a problem which the Wrights had long since solved.*

Cygnet II, *a flying wedge that didn't, was the beloved brainchild of Alexander Graham Bell. It consisted of hundreds of tetrahedral kites (called cells) stacked and joined into a lattice. The engine used in its 1909 test proved underpowered but reports from 1912 indicated that an improved* Cygnet *flew.*

The experimenters of the AEA worked well together. They built a series of picturesquely named "Dromes" (the team's shortened version of Langley's term aerodrome that AGB, in scientific comradeship, had adopted). Drome No. 1 was the *Red Wing,* and Drome No. 2 the *White Wing;* both taught them how to build an airplane and fly it for seconds at a time. The No. 3 Drome, the *June Bug,* catapulted the AEA and Curtiss into world prominence.

It was in the *June Bug* that Curtiss won the *Scientific American* Trophy with a $2,500 prize and $10,000 from the New York Aero Club for flying, on a preannounced date and at a preannounced spot, for more than a kilometer. The AEA picked July 4, 1908, as the date and Hammondsport, New York State, as the place. A large crowd gathered, including a 22-person delegation from the Aero Club. It must be remembered that the Wrights had flown in sight of passersby at Huffman Prairie near Dayton, but not yet at a public event. At Hammondsport, furthermore, a motion-picture camera crew accompanied the bevy of newspaper reporters and photographers.

The tension of the crowd increased as rain and wind delayed Curtiss's attempt to win the prize. The Pleasant Valley Wine Company eased things considerably by providing both food and drink. Around six that evening the *June Bug* was rolled out, and Charles Manly of the Aero Club, formerly with Langley, assisted Curtiss in starting the engine. Curtiss made an attempt at seven but landed short of the goal. Half an hour later he tried again, yellow wings waggling, smoke

pouring from the 40-horsepower Curtiss engine, hands squeezing the wheel in a death grip. At a height of about 20 feet, Curtiss flew for more than a mile. It was not only the first public accomplishment of the feat, but also the first time an aircraft had been filmed by a motion-picture camera and the first moment that Curtiss realized that he would be totally committed to aviation. In a single stroke he had won the *Scientific American* Trophy and had been thrust into direct equality with the Wrights in the world's view.

The flight had enormous impact, not only on Curtiss and the AEA, but also on the Wright brothers, who saw a sudden materialization of the warnings they had received from Octave Chanute about the dangers of competition. They had been asking the Army $100,000 for an aircraft; now they accepted the Army's original offer of $25,000 and began to pursue their European interests more assiduously.

They had shipped an airplane to France in 1907, but in that year had been unable to make suitable arrangements for a demonstration flight. Wilbur, though, crossed to France in the summer of 1908, and at the racecourse at Hunaudières, near Le Mans, began on August 8 a series of flights that astounded the Europeans. He demonstrated to European experimenters what seemed complete mastery of the air, and they quite literally threw themselves at his feet. His flights also catapulted to action a host of competitors who would bring such incredible advances in the next five years that the Wright and indeed all American aircraft would be rendered hopelessly obsolete.

From 1909, the patent fights of the Wright and Curtiss camps had a stagnating effect on aviation in the United States, offset only by the resilience and persistence of a few daring individuals. Neither team distinguished itself. While morally and legally correct in their efforts, the Wrights neglected a national goal—the advancement of aviation as an American strength. For his part, Curtiss became embroiled in some rather unsavory events, the most notorious of which was the reconstruction of the Langley Aerodrome so that it could make flights and perhaps invalidate the Wright patent. Curtiss was also motivated by self-interest to join the ubiquitous Augustus Herring in the formation of a new company at the expense of the Aerial Experiment Association.

It is characteristic of Curtiss that he somehow maintained the friendship and admiration of Bell and company after his breaking away from the AEA, just as it is of the Wrights that they ultimately fell out with genial Octave Chanute. And of course, everyone fell out with Herring.

What a remarkable man Herring must have been to have worked with Langley, Maxim, Chanute, Curtiss, the Wrights and others, and to have enjoyed the admiration of the United States Army and the Aero Club of New York, and yet to have offended all. There is no denying his impact on early aviation—he flew Chanute's gliders, gave Lt. Selfridge advice at the AEA, was with the Wrights at Kitty Hawk, lured Curtiss and others into corporations—and claimed to have in fact been the first to fly. Thomas Selfridge, an acute observer, referred to Augustus Herring as "the first man, so far as we know, to have made a flight in a power-driven machine," apparently believing Herring's claim to have flown an aircraft with a compressed-air engine in 1898 at St. Joseph, Michigan. He also claimed patents—sometimes referred to as patent applications—on methods of control that predated the Wrights, claims sufficiently intoxicating to Curtiss that he joined in the Herring-Curtiss Company on terms extraordinarily advantageous to Herring.

The net effect of the Wright and Curtiss patent battle that began in August 1909 was to send American experimenters searching for ways to beat the patent and to reduce flying in the United States to a series of exhibition teams for whom aircraft of 1908 performance was entirely adequate. Curtiss continued to build aircraft, leasing them to pilots rather than selling them to avoid the legal injunction against sales. He did not hesitate to bring other engineers into his firm. Eventually worn out by the patent fights, and recognizing his own limitations, he lost influence even in his own company. Yet the money flowed in, particularly after the outbreak of war in Europe, and the Curtiss Aeroplane and Engine Company became a valued resource for training pilots and for building maritime patrol planes. Curtiss, however, turned to real estate, and by the end of the war had largely absented himself from aviation development.

In Europe, the events of 1908 touched off a powder keg of achievement. Santos-Dumont had made Europe's first successful powered flight on October 23, 1906, piloting his awkward, banana-shaped *14 bis* fewer than 200 feet. It was more than a hop but far less than Kitty Hawk. A succession of European pioneers hammered out aircraft that could lurch off and onto the ground—Gabriel Voisin, Henri Farman, Louis Blériot, A. V. Roe, Léon Levasseur, Trajan Vuia, Victor Tatin, Captain Ferdinand Ferber, Robert Esnault-Pelterie, William Cody—all names which after 1908 would be writ large.

But only *after* 1908! It was then that the Wrights' insight into the requirement for three-axis control broke the Europeans away from their lemminglike devotion to the idea of automatic stability. Thus the way was made clear for breathtaking achievements during the next five years.

Louis Blériot, for example, had hammered away at the promise of flight since 1902, creating aircraft of almost every configuration, from seaplane gliders to elliptically shaped tandem-winged creations and tail-first pushers, all of which had in common their reluctance to do more than hop from the ground. He poured his own and his wife's fortune into his experiments, and by October 1908 had achieved some slight success, using wingtip ailerons for control. After witnessing

Wilbur Wright conquers Europe from the air. His Flyer, at right, appears in Berlin during 1908. Huge crowds also met him at French airfields; but even in the parking lot coachmen had a good show, below. They steady the horses as the motorized aerial carriage clamors overhead.

Wilbur Wright's enormous triumph at Le Mans on August 8, he had even been quoted in the New York *Herald* as saying, "a new era in mechanical flight has commenced." Further, he unabashedly devoured the details of the Wright wing-warping methods. The Blériot XI, in which Louis Blériot won immortality for his cross-channel flight, used wing warping.

Other European fliers followed suit, and it became a case of "sue and be damned" to the Wrights, because progress was not to be denied. Conditions were far different in Europe than in the United States for the cultivation of flight. In America, distances were great, and the road and rail systems were just coming into their own. There was no practical prospect of an economic return for aviation other than as an entertainment industry. In fact, this is the classification under which the Donald Douglas factory in Los Angeles was listed in 1920. The place itself was an attic above a lumber mill.

Isolationism was the country's political mood—it was conceded that a large Navy was needed to protect the Western Hemisphere from European colonial interests, but the Army was nonexistent, and there were certainly no sizable military funds to be wasted on aviation. By way of contrast, Blériot had made an enormous impression on Europe by flying the channel, and aircraft and airships were beginning to be used routinely in Army and Navy maneuvers and to figure in military doctrine.

There was another difference as well. In America, the aviators tended to be daredevils operating to the advantage of entrepreneurs. The public viewed the Charles Hamiltons and Lincoln Beacheys much as they had the Great Blondin or Harry Houdini. In Europe, there was an aristocratic fascination with flying; it had appeared concurrently with motor racing as a sport, and some of the great racetracks—Brooklands and Le Mans, for example—were converted to aviation fields. There was another important difference as well. In the United States, a professional military officer was not typically wealthy, nor did he routinely have political connections.

France soon took the lead in prewar aviation, providing a homegrown hero—Louis Blériot—who crossed the English Channel on July 25, 1909, in the craft at left. Opposite, his tiny plane sails out from Calais. Above, tumultuous Londoners greet the French aviator, whose arrival meant that Great Britain was no longer protected by her water boundaries.

By 1912, France's sublime Deperdussin broke the 100-mile-per-hour barrier, a signal victory for European technological prowess and a sure sign that America might be losing the momentum generated by the Wright invention.

In Europe, the military was a natural occupation for the sons who did not inherit, and the influence of an officer of the nobility carried weight in quarters where appropriations were made. According to some sources, from 1908 until the outbreak of World War I, Germany had spent more than $73 million on aviation, France $35 million, Russia $34 million—and the United States only $685 thousand. This more than anything else tore the mantle of aviation leadership from the United States.

And it was not as if the new foreign superiority was somehow hidden from the American public. The incredible Reims meet—the "Grande Semaine d'Aviation de la Champagne"—of August 22–29, 1909, had offered huge prizes and attracted 38 aircraft, of which 23 flew. Among the entries, nine were Voisins, four Blériots, four Antoinettes, four Henri Farmans, six Wrights, a Breguet, an R.E.P. and a Curtiss—an update of the *Golden Flyer.* Curtiss won the speed prize at 47.65 miles per hour, narrowly edging out Blériot. But Henri Farman set the distance record of 112 miles at a speed of more than 45 miles per hour, and Hubert Latham in an Antoinette set the altitude record of 508 feet. The message was clear: the airplane was no longer a toy, and America no longer possessed a monopoly.

This became increasingly obvious in 1910, and on America's home turf. A Yankee might well believe he knew how the Europeans had done so well in France—fixing a race was not unknown in the U.S. either. But the Europeans began winning here, in the U.S. of A., at the Los Angeles Aviation Meet, the Harvard-Boston Aero Meet and one at Belmont Park in New York. Worse, those doing it were attractive and glamorous people like Claude Grahame-White, who won the Gordon Bennett Trophy in a Blériot and then proceeded to charm virtually everyone on the East Coast with his easy manners and daring flying. Then there was the veteran Alfred Leblanc, who in his 100-horsepower Blériot had burned up the course—and all of his fuel—to almost beat Grahame-White. Alas, he landed on a telephone pole. And the crowd also responded to the dashing Hubert Latham and the aristocratic Count Jacques de Lesseps, grandson of the builder of the Suez Canal.

By 1911, American aircraft were no longer competitive in European races. The first major long-distance air race started in Paris, ran across France, Belgium, Holland and on to London, then back to Paris. It attracted 43 contestants flying 12 different kinds of aircraft, and only one American, flying a Nieuport! Three aviators were killed, six badly injured, with only nine finalists out of a field of 19 who finished the first leg. The derivative Circuit of Britain attracted 30 competitors, and 12 kinds of English or European aircraft.

While the Wright and Curtiss exhibition teams titillated Americans with sideshow flights of planes with 1908 performance standards, serious money and engineering poured into a profusion of new designs on the other side of the Atlantic. There was also a cross-pollination of engineering and design ideas, a swapping across borders and between companies. Thus the wing warping "borrowed" from the Wrights by Blériot was carried over, with other similarities, into Morane-Saulnier, Nieuport and eventually Fokker aircraft. It was as if the Europeans had agreed by acclamation that protection by patent was not only difficult, but unrealistic as well. One copied, if necessary, but one also improved and changed the design; by 1913 the gorgeous Deperdussin monoplane, with its whirling rotary engine, had gone far past the mythical mark of 100 miles per hour, the "speed barrier" of the first decade of flight. Maurice Prévost, on September 29, 1913, flew his Deperdussin at 126.67 miles per hour.

The final judgment of early aeronautical advancement in the United States might be made on the basis of official records that clearly show the demise of U.S. influence after 1908. In that year, Wilbur Wright held the distance, altitude, duration and speed records documented by the Fédération Aéronautique Internationale. In the years that followed, records were broken in each category every year, but only twice again did an American name appear: during 1910 Glenn Curtiss covered 152 miles on a single flight, and Arch Hoxsey soared to 11,474 feet in a Wright biplane. The rest of the records belonged to the French and to a lesser extent the Germans.

Despite the depressing lid on spending and performance, aviation struggled on in the United States. Europe might be arming itself in the air, Roland Garros might be spanning the Mediterranean, Count Ferdinand von Zeppelin might be conducting regular passenger excursions across Germany in his giant dirigibles, but the exhibition circuit managed to keep alive the spirit that would suddenly blossom after the United States entered World War I.

If the carnival atmosphere of prewar aviation in America

did little to improve airplane technology, it did serve to entertain the crowds that came to watch the aerial heroes—and some heroines—perform their latest, most daring stunts. What names and personalities there were! Harry Atwood, Tom and Casey Baldwin, Lincoln Beachey, the Boland brothers, Walter Brookins, Joseph Cato, the Christofferson brothers, Frank Coffyn, Eugene Ely, Charles Hamilton, Beckwith Havens, Arch Hoxsey and Ralph Johnstone "the Heavenly Twins," the Jannus brothers, the Korn brothers, Ruth Law, Glenn Martin, John and Matilde Moisant, Earle Ovington, Cal Rodgers, Roland Rohlfs, the Stinson family, Dr. Henry Walden, Charles Willard, and so many others!

Each contributed something to history; too often it was a life. Take Lincoln Beachey, for instance. He got his start at 18 in Baldwin airships, learning to fly airplanes in 1910 and exhibiting and test-flying for Glenn Curtiss shortly after that. In those days "daredevil flying" had just become airborne, with as many embellishments in the form of rapid dives and climbs and turns as the pilot dared to make. When the word of Edouard Pégoud's loop reached America in 1913, Beachey had Curtiss build a specially strengthened biplane—the "heavy looper"—for him. It crashed, but the next one was successful; he had added a full repertoire of loops, steep spirals, and long, long dives. Beachey was recognized across the United States as the premier acrobatic pilot, and in 1914 had a brand-new ship built, *The Little Looper,* a pretty biplane with a Gnome engine of 80 horsepower. It was so successful that Beachey soon turned to designer Warren Eaton for an even more radical stunt craft, smaller and faster than before, and a monoplane.

A modern-looking affair with its wingspan of 26 feet and its tricycle landing gear, this even littler looper achieved a top speed of 104 miles per hour. On March 14, 1915, Beachey made his last flight. Taking off from San Francisco's Panama Pacific International Exposition Grounds on a bright Sunday afternoon, Beachey climbed out over the Bay toward Alcatraz. Reversing course he headed into a series of loops, losing altitude with each one. After leveling off and then climbing to 3,500 feet, he dove straight down, finally pushing past the vertical so that the crowd could read "BEACHEY" spelled out on the wings. At a sizzling 180 miles per hour he pulled sharply on the stick to regain level flight, but his wings broke away. Though he survived his plunge into the Bay, the daredevil was trapped in the wreckage and drowned.

Less typical was the brilliant career of the Harvard graduate and race-car driver Charles Willard, who at 26 learned to fly under the direct tutorship of Glenn Curtiss. On August 3, 1909, he made a straightaway flight in the Curtiss *Golden*

Highlights of 1911—cold, tired and angry, Jules Védrines (the Jackie Schmertz of France) won the Paris to Madrid race but was late and the crowds had gone home. He was the only pilot to finish. Opposite, Calbraith Perry Rodgers made the first flight across the U.S. in Vin Fiz, *a Wright EX named for a soft drink sponsor of the feat. During the 84-day trip, he survived five catastrophic crashes before reaching the Pacific. At the very spot, next spring, he flew into a sea gull and fatally crashed. The luck of Harriet Quimby—America's first woman aviator (upper right)—also ran out. She and a friend fell to their deaths after gusts upturned their airplane: they had neither seat-belts nor parachutes. The empty plane landed safely.*

Flyer and almost immediately achieved fame. Fewer than ten days later, with only two hours solo time, Willard broke the newly established, ten-mile distance record established on July 30 by Orville Wright and Lt. Benjamin Foulois, the first U.S. Army airplane pilot. After completing a 12-mile circuit over Long Island, Willard was immediately lionized as the Wizard of the Air. Within months his aerial feats brought him a minimum fee of $1,000 a flight, $7,500 for a ten-day air meet—quite a lot of money in those days!

An exhibit flier who survived could often gather a nest egg large enough to start a business, often in aviation, but the flying business did not always boom. Perhaps it was these hardy veterans of the stunt circuit who coined the aphorism "There's lots of money in aviation—I sure left all mine there." The big money remained in the sky, but aviators were beginning to appreciate the truth behind another of aviation's bywords: "There are plenty of bold fliers and plenty of old fliers but hardly any old, bold flyers."

Willard beat the odds on both counts, going on to a series of firsts: he worked on ground-to-air radio telephones with Lee de Forest, and was shot down near Joplin, Missouri, by an angry farmer who put a bullet through his propeller. Perhaps the first man ever to be brought down by gunfire, he would certainly not be the last.

He was instrumental in developing the arresting gear used by Eugene Ely in his famous, first-ever landing on the deck of a ship, the U.S.S. *Pennsylvania.* With the great good sense to quit exhibition flying, Charles Willard entered a business career that would see him associated with some of the greatest names in aviation.

A Curtiss protégé, Willard joined J. A. D. McCurdy (the first man to fly in Canada) to build a Farman-type pusher biplane. After working as chief engineer for Glenn L. Martin in 1913, he rejoined Curtiss to help develop the big flying boats of World War I. He then formed his own company with Edwin Lowe, Jr., and Robert Fowler in 1915. Called L.W.F., the company featured (quite coincidentally) laminated wood fuselages, the first monocoque-type fuselages built in the U.S. Soon after leaving the firm, he joined Aeromarine Plane and Motor Company.

Willard's later efforts involved less well-known companies, but he was prominent in aviation until his death in 1977 at the age of 94. He was unusual in many ways, perhaps most of all in his longevity despite his perilous early calling.

Many of the other pioneers figured in famous aviation companies. In 1908, at age 19, Joseph Cato submitted plans to the War Department for an aircraft. Later he built several planes of his own design, derivatives of both Curtiss and Blériot types, and then, like so many others, went to work for Captain Thomas Baldwin. By 1916 he had joined L.W.F. and then, in May 1921, he joined the famous Major George Hallet of McCook Field at Dayton to begin long-range development of a powerful radial engine.

"Revolving doors" between industry and the military existed then as they do today, and Cato left McCook Field to start a company with George Elias and Brothers, a Buffalo firm that became noted for a long series of unsuccessful designs. Next Cato went to the Emsco Aircraft Corporation at Downey, California, the site of the present Rockwell International plant. Cato and Charles Rocheville created some beautiful aircraft before the firm folded. In 1941 Cato went back to government service at Castle Air Force Base in California.

Cato and Willard were exceptions. The experience of the Jannus brothers was more typical of the period.

Tony Jannus began flying in 1910, before joining the Benoist Company of St. Louis. Tom Benoist ran a successful flying school and manufactured first-class aircraft. Jannus set a number of records, flying the first airmail ever delivered to Memphis and setting a distance record of 1,973 miles in a Benoist flying boat. Late in 1913, Tony and his brother, Roger, went to Florida to prepare for the inauguration of what is generally considered to be the world's first scheduled flights, those of the Tampa-St.Petersburg Airline.

In 1913, Roger joined Tony at the Benoist plant, where he learned to fly. After helping to establish the run between Tampa and St. Petersburg, the two brothers went into partnership, operating a flying school and doing exhibition work.

After the business broke up, Tony found employment in Russia as the Curtiss Company's Engineering Representative and Acceptance Test pilot. He died when his Curtiss H-7 flying boat crashed and burned near Sevastapol. His brother also rejoined Curtiss, then entered the Air Service. Roger Jannus crashed and burned to death in September 1918, near Champenoise Indre, France.

Even before World War I, the toll of pioneer European flyers was equally grim. Leon Delagrange, Hubert LeBlon, George Chavez, Charles S. Rolls of Rolls-Royce fame, Laffert and Pola, Lemartin, Lieutenant Princetau, Landron,

Gaubert, Gustav Hamel—pioneers all, all their lives important and yet their names are almost unknown. All perished in pursuit of their own personal goals, yet to a man they embodied Otto Lilienthal's famous dictum, *Opfer mussen gebracht werden,* "Sacrifices must be made."

It was but death's beginning; just as Europe wrested the laurels of flight from America, it became embroiled in a great war that nobody wanted. Progress during aviation's first decade, costly as it had been, would be dwarfed by the war-driven advances of the next four years, when the airplane really came into its own.

Of course, well before the first battles of World War I, men were already trying to adapt the airplane to warfare in grim little battles all over the world. Somewhat appropriately, given the current world situation, the first combat flights occurred in Libya in 1911. Captain Carlo Piazza took off in a Nieuport at 6:19 on the morning of October 23, and discovered Turkish troops encamped along the road that led to Tripoli. Several other flights were made that day, some by Captain Riccardo Moizo, who took three rifle bullets through the wing. Piazza and Moizo carried on for the remainder of the campaign, harassing the enemy, attempting to conduct artillery shoots by dropping messages and finally, on November 19, dropping four *Cipelli* bombs on Turkish tents.

The small Italian contingent went on to demonstrate other aspects of aerial warfare besides reconnaissance, artillery direction and bombing. Leaflets were dropped to try to induce the Turks to surrender, and night reconnaissance and bombing flights took place during a full moon. Finally, there were the casualties in action: 2nd Lt. Piero Manzini crashed on takeoff, while Lt. Moizo was forced down behind enemy lines and captured.

America flirted with air power on the Mexican border and in the Philippines, but with inadequate equipment and training too meager to afford success. Yet the very first air-to-air combat occurred in Mexico in late November 1913 between Americans Phil Rader flying for General Huerta and Dean Ivan Lamb for General Carranza. These two exchanged about a dozen pistol shots with inconclusive results.

Similar tentative uses of aircraft occurred in the minor scraps rumbling through the Balkans as Europe tooled up for the Great War. Perhaps more important, major powers of the Old World—France, Germany, Russia and Britain—were actively using their tiny air forces in support of ground maneuvers. The English, with a minuscule army and a tiny air force, demonstrated cooperation between land and air forces so well in 1913 that this approach was key to survival—if not victory—when war came the following year.

A forerunner of turbojet development in the Thirties, the Coanda propellerless plane of 1910 had a tiny compressor fan in its nose; it crashed on its test flight due to pilot inexperience.

Left, Giovanni Caproni worked on early designs, including that of the Ca.9 of 1911, at right. Advanced for its day, the fine wooden monoplane was propelled by an unreliable three-cylinder radial Anzani engine that developed 35 horsepower.

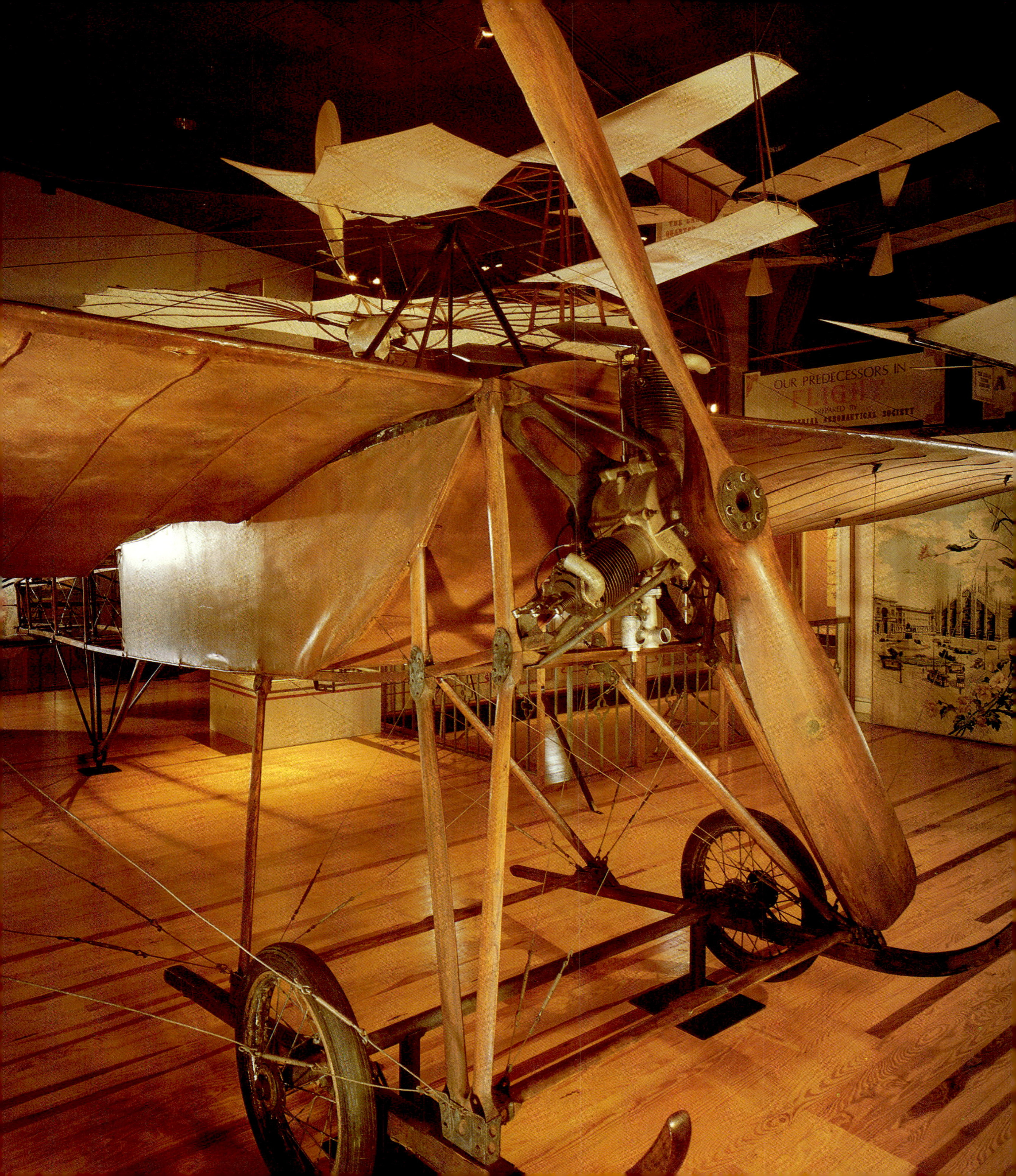
OUR PREDECESSORS IN
FLIGHT
PREPARED BY
AERONAUTICAL SOCIETY

PART 2

BY EDWARDS PARK

The sound was entirely foreign to me. Faint, yet permeating, it wavered into every room of our sunny, country house—a hum without an apparent source. I had been dutifully picking up toys, but the sound had a strange urgency that drove all honorable intention from my five-year-old mind.

"It's an aeroplane!" shouted one of my older brothers, pronouncing the word as carefully as it was spelled in those days. And we erupted from the house, my parents, my sister, two brothers and myself, darting outside to stare. It came over the pine trees, a thousand feet high, yellow wings and dark fuselage startlingly clear against the clean New England sky of 1922.

We gathered in a little knot, drawn together by our wonder. Someone's hands were on my shoulders as we looked up. By then a bit of the astonishment had worn off for my elders, but it was totally new to me, the youngest. I gazed at the airplane's steady flight and savored its great noise—the thunder of mighty cylinders, the slap of wooden propeller blades, the cry of the air riven by struts and wires. It wasn't really a strange new god, just a Curtiss Jenny. But it gathered disciples. Of the four of us siblings, standing there with Mother and Dad, two became pilots and another married one.

Planes gradually appeared more often, and I always dashed out to spot a snarling P-6 or DH-4 or Boeing mailplane. I pored over a thick red volume, *War in the Air,* with its faint, retouched photographs of planes attacking observation balloons, of helmeted pilots waving gloved hands

From 1924, California flier Earl Daugherty steers a Curtiss JN-4 Jenny while his partner, Auggy Redlax, grabs for a cap.

Two Wars and Between

from their cockpits before heading over the lines, of piles of wreckage with natty French officers in *kepis* standing stiffly by.

By second grade, small schoolmates and I talked knowingly of Nieuports and Albatroses, of the way a rotary engine threw oil. With grubby hands we described Immelmann turns and Lufberry circles, and in shrill voices we imitated the blipping sound of a Sopwith Camel coming in for a landing. On small-wheeled bikes, much in vogue in the Twenties, we'd play Richthofen, glued to the tail of the squadron's newest replacement (fresh from Eton), or with nonchalant salute, we'd wheel away from a crippled Rumpler while its rear gunner (poor devil) tried desperately to clear his jammed Spandaus.

We knew all about the great aces. When in 1927, René Fonck set out to fly from New York to Paris, we knew he'd make it. He didn't. He crashed on takeoff. Soon after, handsome, heroic Charles Nungesser and glamorously one-eyed Francois Coli took off from Paris to cross the ocean, and vanished forever.

We youngsters were deeply shocked. We were accustomed to aviation's successes. I didn't remember the NC-4 transatlantic flight, but one of my brothers warned me never to forget that the U.S. Navy beat the British—Alcock and Brown.

When Macready and Kelly flew nonstop across the country, I was al-

lowed to read an account of it in a Sunday supplement which was usually forbidden because of its obsession with passionate crime. I devoured pictures of the Army flyers and their strange single-engine Fokker, windowed like a Pullman car. And I remember my father following the progress of the four Douglas World Cruisers that had set out to be the first to circumnavigate the earth. Two planes made it, and I was told to swell with pride. I did my best.

We ten-year-olds expected this parade of triumphs to continue without missing a step. So when Nungesser and Coli went down, we crashed with them. And when, 12 days later, Charles Lindbergh took off alone from Roosevelt Field, we dreaded another tragedy. I'd gone on an expedition with a brother that May day, and we heard distant bells ringing everywhere, and then a factory whistle. "That guy Lindstrom must have made it to Paris," my brother said.

"Lindbergh," I answered, absently. For the joyous bells stirred echoes in my head. The dream was *not* over. The triumphs *would* continue. The sky was ours, offering a special existence, braver and finer and filled with more splendid rewards than ever a young fellow could reach with his feet on the ground!

So I grew up with aviation, understanding perhaps better than my parents the implications of the midair refueling of the *Question Mark;* or of Wiley Post's pressure suit; of Howard Hughes' sleek H-1 racer, of the giant B-19 which, at my urging, we drove off to glimpse as it flew a planned course, showing itself to America. My generation also learned to accept the dues that aviation exacts, a long list to follow Nungesser and Coli.

When one of my sister's many suitors turned out to own a Waco 10, I knew that she would have to choose him. Fortunately she did. He got me my first ride in a plane—he knew the pilot. When one brother graduated from college in mid-Depression, he knew exactly what to do: go into Marine Corps flight training. We were together in 1939 when the Panzers rolled into Poland, and we agreed that we'd probably get into this new war. "If we do, I'm going to fly," I said. "Of course," he said.

The planes that were born of that war inherited everything we had learned during my growing up. They compressed the sleek lines of the Thompson Trophy racers and the ruggedness of endurance craft. Engines

Opposite, memorial to Allied aviators of World War I was commissioned by a Belgian aviation association. The U.S. Navy's Curtiss NC-4, depicted during a stage of its historic crossing of 1919, above, was the first airplane to fly the Atlantic.

Let's Get'em Flying
5000

Orville Wright at age 72 takes the controls of a Lockheed C-69 Constellation during his last flight. An aircraft of this type, also known as the "Connie," appears at top. During World War II, Boeing employees stand by a special B-17 Flying Fortress that they have just built and signed, their company's 5,000th.

that had inched up in horsepower now surged: 2,000, 2,500, 3,000! The single B-19 that I saw inspired fleets of B-29s; one of these great airplanes was named *Enola Gay*.

One day in 1945, I landed to refuel at a California base, and I heard a noise, a sound totally new to me, insistent, demanding, permeating the whole revetment area. I tumbled out of my cockpit and joined a group of pilots and crew chiefs beside the strip. And we watched a Bell P-59, the first American jet, begin its takeoff. Engines howling, it lumbered forward, then strode, then raced and finally was aloft. We watched it quietly, staying close together as people do when they meet the future.

WWI: The Accelerator

Between 1914 and 1918, war's furious pace forced aviation through a hothouse period of change never again to be matched. During the era, virtually all the possibilities for the development and deployment of air power were explored—strategic bombing, guided missiles, reconnaissance and surveillance and even jet propulsion. This last glimpse of the future was made possible by Henri Coanda's ducted-fan development of 1910 (see page 70). The three types of military aircraft emerged: fighters, observation types and bombers. Such facts as these tempt some military analysts to conclude that, in the final analysis, the First World War was simply a training ground for the Second.

Such a narrow interpretation may offend some of our readers. Yet so revolutionary was the aircraft in the light of traditional warfare—the gentleman's way to fight—that the view introduced above may prove to be a useful adjunct to our thinking about all modern warfare. Also in this category is the fact that during World War I the performance improvement of the aircraft was far less important than the increased percentage of industrial resources and support services devoted to aviation by the belligerents. This wonder of economic organization was founded on the fact that military planners soon realized just how superb the flier could be at spotting targets for the artillery to smash.

British General Sir Horace Smith-Dorrien and his staff salute a B.E.2 biplane during an early air review at Perham Down in May 1913. By the following year, military interest in the potential of aerial reconnaissance had prompted many European countries to experiment with aviation. Detail: an early flight medal, the Imperial and Royal Austro-Hungarian Field Pilot's badge of 1913.

Early in World War I, both German and British recruiting posters lured volunteers with images of chivalrous knights battling evil in the form of dragons. As the war progressed, the horror of trench warfare and mass carnage was difficult to romanticize, so eyes turned upward to the dashing aces who fought intricate dogfights high above the bloody mud of the trenches.

Today we have just about forgotten the role the observation aircraft played in shaping the geography of terror, the trench lines that snaked 400 miles, roughly from Flanders' Fields beside the North Sea southward to the Swiss border. The eyes of the aerial observer, backed up by his nation's big guns, helped to stifle maneuver, and helped soldiers of the line understand their two basic choices—to dig quick or to die quick. Then came the rain, the snow and the blood. Nothing can evoke the horror, waste, carnage, devastation—and futility—of World War I. Even World War II, which was fought over much more of the world, killed more people and at its end produced the prospect of killing everyone, was somehow less horrible than its warm-up war of 1914–1918.

It became impossible for even the most chauvinistic propaganda writer to depict the bloody slaughters of Verdun or the Somme, or French General Nivelle's offensive of 1917 in any positive way. But there had to be heroes and stories for the home front, to disguise the slaughter, palliate the pain and somehow camouflage the muck and stench of rotting flesh in the trenches. Journalists turned to the war in the air for relief: there, killing seemed cleaner. The sky battle recalled the jousts at tourneys of old, and young men in dashing uniforms could be presented to the readers as heroic examples, lures to pull yet more effort, more blood, into the vortex of war.

The big difference between the ground war and the air war was not in pain or horror, but in scale. The death of an 18-year-old boy, fresh out of school, is as sad 10,000 feet in the air as it is in the choking mud of a dugout below trench level. But in scale the difference was vast. There were some 13 million military deaths during World War I from all causes compared to the estimates of fewer than 15,000 airmen killed in all the years and all the theaters of the war. During infamous Bloody April of 1917 (about which more will be said later) 330 British airmen lost their lives; in contrast, 20,000 British ground troops were killed during the first day of the Battle of the Somme in 1916. At the end of that mud-caked 142-day battle, 1,200,000 British, French and German casualties had been incurred. That was a high cost indeed for the exchange of a 30-mile-long sector of land, seven miles wide at its deepest part. Small wonder that stories of the aerial exploits of airmen such as Ball or Guynemer or von Richthofen were more palatable.

A less gory example of scale is that the British Royal Air Force (RAF) at the end of the war was the largest and best equipped in the world, with more than 20,000 aircraft on strength; yet only three percent of the British Expeditionary Force (BEF) was in the RAF. For the other combatants, the percentage was even less.

When the war began in August 1914, neither side really understood the implications of an air war. As the first months of the war sped by, deficiencies of tremendous size and importance were apparent in every aspect of aviation for every country involved.

The aviation mobilization, such as it was at first, was done off the back of each respective general staff's hand; no one had drawn the correct inferences from the very limited warfare that aircraft had engaged in. No one had the slightest idea of the enormous baggage train of equipment, men and supplies, as well as the effort necessary to sustain a single aviation unit in operations.

More importantly, no nation had any idea of the wastage involved in aircraft operations, and none had created the industrial base to sustain active combat operations.

When war broke out, two dozen or more aircraft—a mixed bag of English Blériots, Henri-Farmans, Avros and B.E.2s and B.E.8s landed in France. The first was a B.E.2a piloted by Lt. H. D. Harvey Kelly. He landed at Amiens on August 12 and promptly got in trouble with his commanding officer, who felt that his rank entitled him to be the first to land in France. Operating from outside of Amiens, the tiny force

Men who made history—French mechanic Louis Quénault, far right, used a Hotchkiss machine gun to bring down a German Aviatik aircraft. Sergeant Joseph Frantz, left, flew their pusher Voisin. The October 5, 1914, incident was history's first recorded air-to-air kill.

An observer in a Bristol biplane takes a shot at a German Taube monoplane near Paris, as portrayed, at right, in the Illustrated London News. *When both sides realized the airplane's offensive potential, machine guns became common and dogfights soon developed.*

covered the retreat imposed on the 100,000-man BEF by the huge armies of German invaders. (In World War II, much the same thing happened at Dunkirk.) On August 22, Captain L.E.O. Charlton and Lt. V.H.N. Wadham of No. 3 Squadron observed von Kluck's Second Corps beginning a tactical envelopment of the four divisions of the BEF.

Astonishingly, the British high command believed this report from their fledgling air service. The British began the long retreat from Mons, which injured their pride but saved their army, and it was all based on the report of an aviator. Later that same month, French flyers observed the shift in thrust of von Kluck's armies, and the resulting change in British and French dispositions brought about the Allied victory at the Battle of the Marne that saved Paris.

Thus, while it cannot be said that aircraft won World War I for the Allies, it is fair to note that aircraft prevented its immediate loss.

Sources reporting the numbers of aircraft available for combat duty vary widely, mostly because at the time it was difficult to define what a combat aircraft was. Most analysts report that the Germans had about 450 aircraft, a little more than half of which were ready for combat duty. The Germans could also muster 11 dirigible airships. The French could field about 300 first-line aircraft from an overall strength of 600 and about a dozen dirigibles. The British had, at home and in France, about 160 aircraft, of which perhaps a third were "front-line quality." Britain never really came to grips with the dirigible question until after the war.

In Germany, as elsewhere, there was a multitude of types, ranging from the graceful Taube monoplanes to the very efficient Albatros biplanes designed by Ernst Heinkel, later to be known for his World War II designs, including the first operational jet aircraft. German aircraft tended to depend upon the heavy but reliable six-cylinder, "automobile" type of water-cooled engines of 100 horsepower. The Germans liked strong, fast machines and appreciated streamlining.

Both Britain and France placed equal dependence on rotary and stationary engines, and through the war both countries manufactured scores of variations of each type. Both liked agile machines with a good rate of climb. Considering that no engine was produced to a standard of parts interchangeability, a goal of American production, the multiplicity of types posed a spares and maintenance problem that was almost insoluble.

On the Eastern front, neither Russia nor Austria-Hungary could initially deploy an air force of any weight, nor was there an industrial base to bring one to the field. The Russians built copies of Voisin and Albatros aircraft and had purchased examples of several other types. Their initial—and perhaps greatest—contribution to the air war was the creation of the

Grooves on the armored wedges bolted to the propeller of Jean Navarre's Morane-Saulnier N monoplane, right, enabled the French ace to shoot his 8-mm Hotchkiss machine gun with relative safety through the disk formed by his whirling propeller. To aim his gun he aimed his plane—a great advantage in a dogfight and a technological advance that would forever change the face of aerial warfare.

Another creative application of air power included resupply of ground-based units, above: two U.S. Marine pilots with the Royal Air Force Squadron 218 dropped more than a ton of supplies in flights to an isolated French infantry regiment in October 1918. Overleaf: smoke trails from a stricken German aircraft in a cloudscape over the Verdun Front—a clear victory for the French Nieuport 11.

first four-engine bomber, Igor Sikorsky's *Ilya Muromets*. More than 70 of these large aircraft were built in the course of the war, conducting extremely successful operations until the Russian collapse in 1917. Austria-Hungary attempted to create an air force under free-market conditions, and the result was utter chaos in design, supply and operation. Indigenous Austrian designs proved unsatisfactory, and German types were purchased in order to have some capability. In general, the air war on the Eastern front could be considered unimportant until well into 1916. The Italians possessed a fledgling force with close to 100 planes, most of foreign make, but by 1918 some good fighters would appear from domestic factories.

Like many other discoveries during World War I, the perception of the value of aircraft developed as a result of miscalculation. The one situation that no nation wanted—a static war—developed. Trench warfare with its massed machine guns spelled the end of cavalry scouts, and aircraft replaced horses for reconnaissance, permitting for the first time the dream of every general: to be able to "see what was happening on the other side of the hill." The Germans, engaged with huge Russian armies in the East, were short of manpower for the Western Front. The Germans relied on heavy artillery for defense; aircraft were used to register the cannon fire. The evolution of trench warfare thus dictated the initial task for aircraft; in turn, their success made trench warfare more permanent.

Thus the air war unfolded. If the principal work of aircraft was reconnaissance and artillery spotting, an equally important task was the prevention of the other side from doing so. Prevention required "scout" aircraft, as "pursuits" were first known. We now call them fighters. Ideally, the scout attacks would be foiled by other scouts; thus a new scenario was established. Inevitably, the war started with unarmed aircraft, most often observation two-seaters; it ended with a predominance of single-seat fighters intent on first eliminating each other and then eliminating the two-seaters.

By 1915, aerial reconnaissance, both by eye and by camera, had reached such a level of sophistication that all of the maps for the Battle of Neuve-Chapelle, which began in March of that year, were based on aerial photos. For the first time, bombing raids in direct support of ground operations were conducted in an attempt to destroy the railheads through which German reinforcements had to come.

More importantly, all armies began to realize that the support of aircraft in the field required a sophisticated and expensive baggage train that reached not only back to the factories in England, France or Germany but well beyond to the purveyors of raw materials. In the past the aviation industry had existed in the shadow of general armaments. Its needs had been more than adequately supplied by civil production.

While aerial reconnaissance remained important, the machine gun, demonstrated above by its inventor Sir Hiram Maxim, changed not only the pattern of ground warfare but also helped transform the airplane from a passive observation craft into an active weapon of war. An aerial view, opposite, of the French infantry after a bombardment on the Somme.

Now requisitions were being placed for wood of exceptional quality, for steel suitable for high-powered, lightweight engines, and for essential materials such as castor oil, rubber, copper, brass and linen, all in quantities never before imagined. The price of everything connected with aviation was high: the time involved in training mechanics and pilots, the creation of the increasingly specialized engines and aircraft and the pay of troops in the field. An aircraft mechanic in the British service earned roughly seven times the pay of a foot soldier, with a similar premium paid in other air services. General staffs that had calculated the cost of fighting in terms of iron rations, fodder and cannon shells were now faced with equipping squadrons with equipment of Rolls-Royce quality.

Machine guns had been mounted on aircraft almost from the start of the war; the effect of the extra weight was so detrimental to flight performance that these arms were immediately removed. But everyone learned. Different aircraft came to the front, and they were systematically tuned for a gain in performance. The concept of height as conferring high speed in a dive and serving as a store of energy came into being early; it is still valid today, bringing with it a distinct combat advantage.

The first step toward the development of real pursuit or fighter types took place when Roland Garros of French Escadrille M.S. 23 arranged for bullet deflection devices invented by Raymond Saulnier to be attached to the propeller blades of his Morane-Saulnier Type N in March 1915. The Type N had a Le Rhône rotary engine of 110 horsepower and a top speed of 103 miles per hour at 6,500 feet. The performance was not remarkable, but the bullet deflection wedges attached to the propeller blades of Garros' Morane-Saulnier permitted him to fire a Hotchkiss machine gun through the arc, or "disk," formed by the whirling prop without shooting off the blades.

Garros did not need to be a spectacular marksman; he simply needed to aim his plane and pull the trigger. The effect was revolutionary. Garros shot down a two-seater on April 1, 1915, and brought his total to three in the same month. He scored again on April 13 and April 18, but on April 19 a single rifle bullet fired from the Mauser of an aged, surprised German reservist near Courtrai cut the Morane's gasoline line. Garros had to force land, and the secret of his success was out as he was unable to destroy the aircraft.

Copying the deflectors was easy; making the devices work was not. Either the metal the Germans used on the deflector blades was inferior, or the armor-piercing quality of the German bullets was superior. In German experiments, the gun simply shot the propeller off. As a result they sent—as the German High Command so reluctantly yet so often had to do—for the young Dutch designer Anthony Fokker.

Only 25, a man whose real genius would constantly be tarnished by bragging and problems of quality control, Fokker rejected the primitive deflection system and directed his staff to create a synchronizing system for the Parabellum machine gun to be mounted on his new Fokker M 5K monoplane. The synchronizing system developed by the Fokker team momentarily interrupted the fire of the machine gun whenever the propeller blades were in front of the muzzle; therefore no bullets could hit the blades except in a hang fire. Eventually the Allies would develop their own device, the Constantinesque interrupter gear, but not before Fokker's device gave Germany a significant advantage. Curiously, some sources report that such a device had been designed in France before the war but was not employed.

Fokker's M 5K was derived from an earlier Morane-Saulnier; both owed much to the Blériot XI, also a French machine. The M 5K had an 80-horsepower Oberursel rotary engine copied from the French Gnome, and a top speed of about 82 miles per hour. It sufficed; with the synchronized

Glorified at home while earning the grudging respect of their brothers in the trenches, World War I aces seemed to have it all. In fact, ill-suited equipment, poor training and no parachutes helped to create high mortality rates—half of the aces pictured here survived. The *Order Pour le Mérite* was the highest honor a German aviator could earn. It was dubbed the Blue Max by the British after Max Immelmann, center, a brilliant German ace and aerial tactician. Prominent French ace Charles Nungesser, center, flouted death continually, suffering 17 wounds and injuries during the war. He ignored the superstition of many fellow aviators with his characteristic skull, crossbones-and-coffin symbol. France's ace of aces René Fonck, opposite, relied on skilled marksmanship and brilliant airmanship to earn his 75 confirmed victories. Best known of all World War I aces, Germany's Manfred Freiherr von Richthofen—the "Red Baron"—stands for inspection with the German unit Jagdstaffel (Fighter Group) No. 11 in early 1917, one month before the successful campaign against Allied air forces known as Bloody April. Squadron affiliations such as distinctive insignia cemented bonds between aviators: right, Sergeant Raoul Lufbery stands with his Nieuport 17 bearing the headdressed Indian symbol of the Lafayette Escadrille, a squadron composed of American volunteers who fought with the French before American entry into World War I. Right, leading American ace Captain Eddie Rickenbacker and members of the 94th Aero Squadron lean against a French SPAD bearing the group's colorful "hat-in-the-ring" symbol.

Eddie Rickenbacker (center)

Raoul Lufbery

Max Immelmann

Manfred Freiherr von Richthofen (center)

Charles Nungesser

René Fonck

Leergewicht: 2740 kg.
Nutzlast: 1235 kg.
zul. Gesamtgewicht: 3975 kg.
Auf rechter Seite aufsteigen.

Built in 1924, the British Vickers Virginia, below, and its tremendous arsenal provided a postwar answer to the Gotha series. Largest of all the World War I bombers, the German Zeppelin Staaken VGO III, bottom, had a wingspan of 140 feet, just three feet shorter than the wingspan of the B-29 of World War II. The seven-man crew included a mechanic and two pilots, the latter to work unwieldy controls of the giant aircraft.

Protected by a nose gunner who, with two more crew members, sucked on oxygen tubes during the high-altitude bombing sorties, Germany's Gotha G V bombers brought a new dimension to destruction: their raid in June of 1917 caused 600 British casualties. Germany's heavy bombers first attacked Britain in late 1916, after the German high command halted ineffective attacks by Zeppelins.

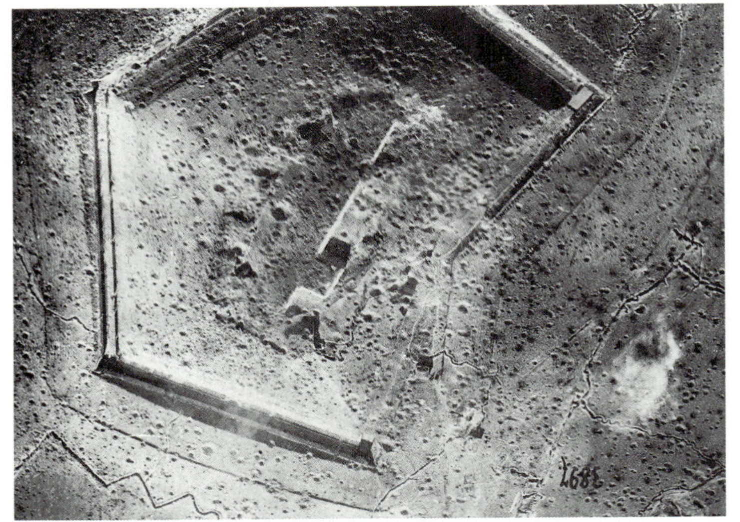

gun, the tiny single-seat monoplane with its wingspan of 28 feet became a world-beater and winner of medals or "gongs," as the decorations of the day were called.

The German High Command (OHL) ordered 50—to be known as Eindeckers—for immediate delivery. Parceled out to various squadrons as they were built, they became the object of much desire of gong hunters. In the hands of Lts. Max Immelmann and Oswald Boelcke, the Eindeckers had become known as the "Fokker Scourge" by mid-1915. They vastly affected not only the air war on the Western Front, but the subsequent development of aircraft.

The actual casualties caused by the Fokkers were relatively few; far fewer than the daily attrition ("ordinary wastage" is the callous term of the day), far less than individual snipers caused, nothing compared to the cost of local eruptions of artillery fire, not a jot compared to casualties from trench foot and other related iniquities of the troglodytic life in the line, and certainly negligible compared to the casualties from starvation on the Eastern Front. Yet the losses were profoundly unsettling to Allied airmen who, perhaps far more clear-sighted than the generals they fought for, peered into the unknown future.

Opposition to the Fokkers was light at first; the British were flying the stable, slow B.E.2cs and Maurice Farman "Box Kites" and sought some sort of mutual protection in formation flying. The French employed Morane-Saulnier Parasols, "military" Blériots, Farmans and Voisins. Their Caudron bombers were soon retired to night duty after heavy

German aerial photographs, left, reveal the course of devastation at French Fort Douaumont in February 1916. A defensive fortification on the outskirts of Verdun, the stronghold resisted heavy shelling at the start of the Battle of Verdun, during which 700,000 French and German soldiers died.

losses during day raids. None was capable of standing up to the Fokkers, whose highbred pilots would stalk the front as if they were hunting stag in Romitern Forest.

Riposte to the "Fokker Scourge" came quickly. Captain Geoffrey de Havilland designed the F.E.2b and D.H.2, both pusher types, to obviate the need for a synchronizing gear. The F.E.2b had an enormous, inefficient Beardmore engine, and in addition to being overweight, underpowered and unstreamlined, the airplane possessed a top speed of only 81 miles per hour. The D.H.2, capable of 90 miles per hour and very maneuverable, mounted a fixed Lewis gun in front of the pilot and soon proved itself to be the Fokkers' master. Another noteworthy aircraft, the Vickers FB.5 Gun Bus, forecast the future trend of the war by being the first type sent into action in an homogenous unit, No. 11 Squadron, R.F.C. (Royal Flying Corps). The squadron went off to France on July 25, 1915.

The French had responded to the Fokker threat with the lovely Nieuport 11 or *Bébé* which was very maneuverable and fast at 97 miles an hour and armed with a Lewis gun mounted on the top wing to clear the propeller disk. It worked with the D.H.2 and the Gun Bus to eliminate the Fokker threat; in doing so the Allied craft called in their own 1916 nemesis, the German Albatros, perhaps the first airplane ever planned from propeller to rudder as a pure fighter.

As the air war spread, airplanes became significant even when used in small numbers. In 1915, British Short seaplanes began operations from H.M.S. *Ark Royal* against the Turks in

Aiming at infantry below, a British bombardier of early World War I prepares to drop a bomb. Ground-based units such as these German troops, left, quickly learned to protect themselves with anti-aircraft guns. Aerial shell bursts were termed "archie" by Allied pilots after a popular prewar song.

the Dardanelles. The Italians broke their alliance with Germany and entered the war on the side of the Allies on May 24, 1915, with a strength of just under 100 aircraft. They proved not to need the farsighted tactical and strategic teachings of General Giulio Douhet, who had initiated aerial warfare in Tripoli in 1911. In the "Big War," the Italians went on to engage the Austrians in pinprick air battles, primarily with seaplanes.

In contrast, the Germans found that their fighters (and other aircraft, for that matter) were more effective if grouped in homogeneous units, and while they never achieved the consistency of type assignment found in Allied units, the groundwork was well laid for the development of the later Jagdstaffeln (fighter groups). More than 80 individual sections of aircraft were deployed by the Germans by the end of 1915, including specially designated bombing squadrons. Such expansion demanded the creation of a chain of flying schools, training courses for observers and mechanics and the establishment of tables or organizations of great size. The tail had begun to wag the dog; by 1916, a 14-plane fighter group required 117 people to support it.

The British and the French went through a similar evolutionary process, tailored to their national military organizations and national resources and, most important, grounded in their respective approaches to warfare. Thus, deployment doctrine emerged for the aerial service.

In view of the German occupation of parts of France and Belgium, British war doctrine called for an unremitting offensive military posture to force the Germans back to their own soil. In accordance with this directive, British air chief General Hugh Montague Trenchard insisted that the air war be carried to the Germans under all circumstances, in spite of the relative merits of opposing aircraft, in spite of the survival statistics of pilots on the front.

The French maintained a middle course, fighting aggressively when required, maintaining the status quo when there was not an important ground engagement to support.

For their part, the Germans husbanded their resources, making no attempt to do more than oppose the Allied aerial incursions that they perceived as threats. Germany possessed fewer aircraft, but its generals systematically deployed them in concentrations that could win local air superiority. German fighters depended less on "patrols" than did their Allied opponents: instead, they relied upon an increasingly sophisticated warning system to analyze threats before the pilots launched their aircraft. These tactics were eminently sensible,

Three men who characterized German aerial prowess: from the left, Bruno Loerzer, head of Jasta 26 and expert flier-fighter with 41 victories; Anthony Fokker, Dutch designer of the highly successful Fokker D VII and other machines of war; Hermann Göring, leader after the death of von Richthofen of Germany's most feared pursuit squadron, the Jagdgeschwader 1.

particularly given the prevailing winds, which blew toward the German lines.

German air superiority was at an end by June 18, 1916, when the plane of Max Immelmann broke apart; credit was given to an F.E.2b. After the death of their "Eagle of Lille," the Germans began to group their fighters into squadrons as they tried to adjust to the flood of French and British aircraft. The lesson was driven home during the infamous Battle of the Somme. British and French aircraft outnumbered the Germans three to one at the start, and two to one by the finish.

The big shake-up in 1916 of the German High Command, which saw Feldmarschall Paul von Hindenberg and General Erich Ludendorff take over effective control of not only the army but the state, resulted in a serious new production program for aircraft. But the equipment available to the Germans—Fokker and Halberstadt fighters—was inadequate, and in desperation the High Command called for copies of the Nieuport to be made. Some companies produced almost identical aircraft; the Siemens-Schuckert D I was a virtual duplicate of the French biplane. As so often happens, one of those imitations turned out to be superior to the original, the Nieuport in this instance. In the hands of German ace Oswald Boelcke, the Albatros D I would help write many tactics of air warfare that remain in practice today.

The inspired imitation—a sleek, shark-nosed fighter by designer Robert Thelen—was not so maneuverable as the Nieuports, and only slightly faster, but it mounted two Spandau machine guns. With weapons double the firepower of any on the Allied planes, Boelcke established Jagdstaffel 2, coaching his men rigorously in his own rules for air warfare. Boelcke's "cubs" included Manfred von Richthofen, the future "Red Baron" of Allied dread. All were carefully drilled before he launched them into combat in the fall of 1916. Boelcke himself was to be successful in 40 victories before colliding in midair with one of his protégés, Lt. Erwin Böhme, on October 28, 1916. With Boelcke's death the Germans began their solemn deification of fallen air heroes, which was to extend through the end of World War II.

In the hands of Boelcke's students, the Albatros D I gained aerial superiority for Germany for a time. At its introduction, the Albatros was clearly superior to all Allied fighters and was to predominate in the German Air Service through the summer of 1918 and to be in use until the end of the war.

It was with the Albatros that the Germans, spearheaded by the legendary von Richthofen, conducted the 1917 aerial campaign dubbed Bloody April. Sitting behind two reliable

German air superiority during early 1916 could be attributed to the interrupter gear, left, a device designed by Dutch manufacturer Anthony Fokker and his team. By synchronizing machine gun fire with the spinning propeller, Fokker's invention enabled German aviators to fire directly ahead, assuring improved accuracy and more kills.

Evolution of Fighter Tactics: WWI

Early in World War I, there were no true fighter aircraft or air-combat tactics. The first real fighter arrived in 1915 in the form of the German Fokker Eindecker, which introduced the machine-gun synchronizer. This device allowed the pilot to aim a fixed machine gun by pointing the entire aircraft, and to fire forward through the spinning propeller.

The early fighter pilots were usually "lone wolves," who stalked their prey, often diving out of the sun or attacking from the victim's blind or undefended quarter. Then they would either dive away to escape, or "zoom" climb, regaining altitude and safety while awaiting another attack opportunity. One of the most famous of these tactics was the "Immelmann Turn," named for one of the original Fokker pilots.

Gradually, defensive formations arose to combat these deadly tactics, and inevitably fighter formations grew larger so that strength could be met with strength. By the end of the war, these "Flying Circuses" often numbered about 50 aircraft. Since communications between fighters were normally limited to only visual signals, however, coordination was difficult. Once the fight began, it was every man for himself.

ROBERT L. SHAW

The Immelmann Turn—*(World War I) Western Front near Cambrai, France, 1915: spotting a British B.E.2 biplane below him, a German Fokker Eindecker dives to attack (upper left). The B.E.2 evades by turning (lower right), and rather than waste a machine gun burst the Fokker climbs away (upper right). Executing a "rudder reverse," the Fokker spins vertically into a second dive, with a clear shot at the British airplane.*

machine guns, they were able to establish—for the last time—German air superiority over the Western Front. The Germans first concentrated against the British when the French, battered over Verdun, had withdrawn much of their air service for refitting. The Germans inflicted casualties at such a rate that the expected flying life for Royal Flying Corps pilots was officially calculated at 17.5 hours. Vanquished British planes numbered 151, against the loss of 51 for the Germans. Richthofen ran his string of victories to more than 50 during this period, and even Trenchard began to waver in his commitment to the offensive.

The Allies, however, were quick to respond with both new aircraft and better training and tactics. Schools for "special flying" had been established in Britain in May 1916. Here veterans back from the front passed on the facts of life and death to newly fledged pilots. The tactics they evolved were not very different from the *Dicta Boelcke* though they emphasized the offensive spirit.

The composition of Allied units began to be dictated increasingly by the resources available to the individual powers. An embarrassing multiplicity of aircraft types and engines began to bloom on the Allied side, creating chaotic supply and maintenance problems, while shortages forced the Germans to concentrate on a few basic makes of engines and designs of aircraft suited to them. In general, the Allies produced engines of superior horsepower. Both sides produced rotary engines throughout the war though it was clear by 1917 that this form of prime mover had reached the end of its development road.

Short of rubber, fuel, brass, copper and other essentials of aircraft making, the Germans generally had to make do with engines of lower horsepower. Beginning in 1917 and continuing until the end of the war, German airplanes typically mounted the 160-horsepower Mercedes or 185-horsepower BMW engines, while their Allied counterparts might be powered by a Hispano with 235 horsepower, a Rolls-Royce Eagle with 375, or even an American Liberty engine of 400 horsepower.

The German aeronautical engineers overcame their power limitations in some measure by tailoring the aircraft to the available engine and to the mission. But a good many Allied fighters exceeded the speed and altitude capability of the opposition. Allied two-seaters, including the Bristol fighter

Fearing meager volunteer turnout for the Air Service, the U.S. War Department distributed patriotic posters, announced financial incentives and promised high rank. Fears proved groundless: the Air Service was flooded with eager applicants for flight training.

When the United States mobilized for war in 1917, an American who served in the Royal Flying Corps, Captain Geoffrey Bonnell (right) and an unidentified Lieutenant, visited New York to orchestrate the recruiting program for the British air forces. Americans serving with the British often flew S.E.5as, as seen in Squadron No. 85, opposite.

or the de Havilland D.H.4, functioned in several missions. German two-seaters, mostly observation types, were carefully crafted to extract all the performance possible from the design for a single mission. By accepting numerical inferiority and by determining that they would fly only essential missions, the Germans remained competitive until the final bitter days of late 1918.

To carry the air war to the Allies, the Germans relied on their Zeppelin airships and various bombers, which they used to conduct raids against targets in Britain. The first instrument of frightfulness, almost 90 Zeppelins were built during the war. Squadrons entered England's airspace, but only sporadically and with little material effect. From the first three-ship Zeppelin raid of January 19 and 20, 1915, when 24 bombs were dropped on Norfolk, killing four persons, to the last five-ship raid on August 5 and 6, 1918, damage worth only £2 million was done, hardly even worth the loss of the famed airship commander Korvettenkapitän Peter Strasser during this final incursion. Far more effective was the initial attack on May 25, 1917, over Folkestone, without opposition, of 21 twin-engine Gotha bombers. Britons were outraged at the result—95 dead and 195 wounded.

Their classroom formation, the audacity of bombing in daylight and destruction of *private property* caused a mini-revolution. Britons demanded protection. Germans accelerated the pace; action even on the German home front was heroic—given the difficulties in building heavy bombers. On June 13, 1917, bombs fell on London, killing 162 and injuring 438. But the hurt went deeper: something had to be done. The British responded by creating an efficient system of aircraft reporting, antiaircraft defenses and night fighters that served as the basis for German defeat in the air over England in both world wars.

The Gotha raids had another impact on the future. The opposing forces drew opposite conclusions from the effort. Convinced by the ratio of resource expenditure to result, the Germans decided that strategic bombing was not the best use of armed forces. World War II's reborn Luftwaffe found itself geared primarily to army support. Determined that they should not only get angry but also get even, from the mid-Thirties the British made up their minds to create a bomber force that would carry the war to the German heartland.

Observation ships and doctrine also evolved rapidly on both sides. The task of artillery spotting had developed into a fine art with the use of the "clock system" of spotting shell-fire. Aerial reconnaissance and surveillance aircraft emerged, some capable of long flights at altitudes of more than 20,000 feet. Both sides routinely photographed the entire front twice a day. Ground attack came into vogue on both sides; fighters were used at first, but their vulnerability to the hornet's nest

C1904
D
6851
C1931
B
7870

of machine-gun fire from the ground—both friendly and unfriendly—led to the development of special armored types. At sea, Allied flying boats established antisubmarine patrols, and twin-float fighters were developed to oppose them. The sea war went from the Baltic in a great arc around Europe and into the Black Sea; it was extended even farther when the Germans equipped ships with scout planes for reconnaissance.

All combatants bombed military targets by day and by night, attacking railheads, ammunition dumps, airfields; both sides accused the other of attacking "plainly marked hospitals." Even in bombing, the propagandists had seen advantages.

The Allies wrested air superiority from the Germans once and for all in the fall of 1917, when a series of improved aircraft (often dangerous to fly) reached the front. That "fierce little rasper," the Sopwith Camel, was to kill more Germans—and more Englishmen—than any other airplane. Another was the S.E.5a, a handsome, conservative-looking airplane that pilots loved because of its forgiving flying qualities and stability as a gun platform. The Bristol fighter, the "Brisfit," which had made an inauspicious debut when Manfred von Richthofen's unit shot down four out of the first six on a patrol, came into its own. Molded into one of the most effective weapons of the war, it remained in British service through 1932. The French fielded the SPAD XIII: it outperformed the SPAD VII but because of engine problems was far less reliable.

In 1918, Fokker produced the aircraft that kept Germany competitive. An intuitive pilot who hired designers of genius, Fokker had in 1917 introduced the Dr. 1 Triplane, the type in which Manfred von Richthofen won a number of victories and in which he was ultimately killed. As agile as the Sopwith Camel, the Fokker Triplane arrived in relatively small quantities (only about 170 at the front during May 1917) and served for only a year. Despite quality control and structural problems, it rehabilitated Fokker's image; he was invited to compete in the first great fighter competition at Johannisthal, near Berlin, in January 1918. He offered the Fokker D VII, a craft which represented a true departure from all previous aircraft designs of World War I and which set the standard for the next decade of aircraft development.

It was an angularly handsome biplane, with construction features typical of its designer, Reinhold Platz. The thick, almost-cantilever wings were of wood and fabric, with Fokker's customary welded steel-tube construction for the fuselage. Powered by the ubiquitous 160-horsepower Mercedes, the Fokker D VII had a top speed of only about 120 miles per hour and a ceiling of 23,000 feet. But it was maneuverable and, most of all, forgiving; it was said that it made good pilots out of bad ones and aces out of good ones.

The D VII became the standard fighter of the German squadrons for the last of the war. It was the only aircraft specified by name in the Armistice Agreement to be turned over to the victorious allies—*In erster Linie alle apparate D VII* (especially all first-line D VII aircraft). As fine an airplane as it was, the Fokker D VII could do little to turn the Allied aerial tide that crested over Germany's armed forces. That

America gears up for war: above, workmen polish propellers; right, Packard Motor Company workers test the Liberty engine, a 400-horsepower V-12 from Detroit automobile manufacturers. Liberties powered the British-designed, American-built de Havilland DH-4 two-seater, the only World War I combat plane made in America, opposite.

Muddy French training fields, opposite, were the first stop for most fresh American pilots such as Lt. Reynolds Benson, right. He might have flown a Curtiss Jenny at home, but nearer the front was introduced to a variety of French planes. In any unit such as the 94th Aero Squadron, above, the vast majority of personnel were not pilots but support staff—mechanics and other ground crew members who kept the relatively few pilots in the air.

nation's fate was sealed when America joined the combat in April 1917; its greatest contribution to the air war unquestionably was its thousands of enthusiastic pilots.

The story of America's material contribution is usually depicted as a colossal boondoggle in which hundreds of millions of dollars were wasted in an effort that ultimately got only a few hundred combat aircraft in the field. The Americans, though, entered the war with a handful of obsolete aircraft and without an aircraft industry or even an industrial base upon which to build one. In the space of a year the country was transformed: armies of lumberjacks selected prime spruce, farmers planted acres of castor beans for lubricating oil; industrialists not only created factories for fittings, linen, dope, engines, instruments, wire, armament and all of the thousands of other items necessary for a combat airplane, but also trained the tens of thousands of workers to make them. Aircraft manufacturers seized upon existing designs in 1917, and were building at the rate of 12,000 or more aircraft per year by the end of the war in November 1918.

Though not a war-winning weapon by itself, clearly the aircraft was an indispensable weapon, just beginning to reach the front in numbers necessary to be really effective. U.S. Brigadier General William "Billy" Mitchell's use of the American and French air forces at St. Mihiel, Trenchard's development of the Independent Force of bombers and the British and even the German air defense system development were all the top tier of that mass employment. Underneath were the special ground-attack aircraft employed by Germany in precise tactical use (bombing and strafing), increasingly sophisticated use of large formations by both sides and, of course, the expanding pyramid of support that ran from fully equipped frontline squadrons back through depots of spares, test centers and an immense manufacturing network.

And yet the stage had merely been set. Air power in World War I was to air power in World War II as sea power at Trafalgar was to sea power at Leyte Gulf. The measure of the war's advancement of aviation can be noted both in its final standard products as well as in its aberrations.

In terms of standard products, the Fokker D VII could hardly be improved upon, but it was closely matched on the Allied side by the British S.E. 5a, the Sopwith Snipe and Dolphin and the prospect of the Nieuport 29 from the French. The German bombers had become notorious, but their English equivalents, the Handley Page 0/100 and 0/400, were better airplanes. The Italian success with Caproni bombers has been mentioned, but Italy also developed the excellent Ansaldo series of fighters and reconnaissance planes. In two-seaters, there were remarkable and often unremarked-upon successes, particularly by the Germans. The Albatros two-seaters were almost uniformly successful, as were the A.E.G.s, D.F.W.s, L.F.G.s, L.V.G.s and others. The Halberstadt and Hannoveraner ground-attack airplanes were supplemented by the almost indestructible Junkers J-1 armored biplane. The British Bristol F.2B became beloved by its crews and feared by the enemy, and even de Havilland's maligned D.H.4 and D.H.9 aircraft did good work.

The French produced a number of excellent two-seaters, including the Breguet 14, which sired a generation of able descendants, and the Salmson 2, which was welcomed by American units. The French also produced some elegant multiplace twin-engine aircraft, including the Caudron R. 11, which could have served until the Thirties in any air force in the world without remark. War spawned enough resources for the dreamers and the eccentrics to flourish. On the English side there were such ponderous creations as the Pemberton-Billing P.B. 29 with four wings, which would have shot down Zeppelins if it had been able to catch up to one. Other false starts included the twin-fuselage seaplane, the A.D. Seaplane Type 1000, termed "a floating conservatory" for its many-paned windows by Air Chief Marshal Sir Arthur

Unsung heroes of the air war, mechanics of the American Air Service, overhaul the airframes of a Nieuport, left, and a Breguet, right, in a repair shop.

Flying the SPAD XIII

A. Raymond Brooks

The SPAD Pursuit (Fighter) single-seater biplane was a "flying brick" as my first flight proved to me on June 5, 1918. But it was joy to manuever it in subsequent dogfights because you felt completely confident that this bird had the required 100 percent agility and power and strength.

My basic training was in Texas on the famous Curtiss JN–4D with OX5 90-horsepower motor; the advanced was in Issoudun, France (3rd Aviation Instruction Center) on Nieuport's 23, 18, 15. These planes "floated" (glided) noteworthy kilometers to safe landings in event of motor failure. The SPAD, however, had to be flown to the landing area with gentle power—or else, a crash. So, on my first try on a type VII with 180 horsepower Hispano-Suiza water-cooled engine, my learning was done the hard way. At Vaucouleurs south of the city of Toul, the aerodrome was on a wooded stretch high above the town itself.

With a roomy cockpit and buckled into the new safety belt having a single button for release of four chest-positioned straps, it was simple to yell "coupe" ("cut" on two magneto switches) and to work the throttle to the proper "contact" moment with the mechanic swinging the propeller. Takeoff with 2,000 rpms was normal, and I just had fun gyrating over the valley and testing the controls in various attitudes and convolutions. I was impressed strongly by this little ship's tendency to drop like a stone when motor power was reduced. My concern to make a perfect touch of the landing-wheels to the turf was so great that I made four passes before I had safe forward speed combined with very little altitude to cut the power. (No brakes, on any of our machines.)

Overleaf: As if ready for one last mission, Capt. Brooks' SPAD, magnificently restored by Smithsonian artisans, stands in the Institution's National Air and Space Museum.

Our enthusiasm (or *esprit de corps*) was enhanced July 31, 1918, by deliveries of SPAD XIIIs with 220 horsepower, geared "Hisso" and *two* Vickers or Browning guns. All these elements improved our chances of survival. The SPAD XIII fighter of WWI had all that I could wish for in combat. On climbing to 18,000 feet—which took about 20 minutes—the lack of oxygen made the mission increasingly perplexing. Our flight "monkey-suits" were of canvas lined with dog hair, dyed with something that soiled our uniforms. Our big, well-insulated mittens had a large pouch for the fingers that was partly removable for fore and middle fingers to operate two triggers vertically mounted below the top of the "joy stick" control. The rudder bars had wire loops like croquet wickets over the thick foot-moccasins encasing our boots. I once used the loops myself when one rudder control wire was bullet-severed.

And our ordnance! Our air-cooled machine guns were jamless if we meticulously checked our tracer, incendiary and armor-piercing projectiles for true diameter before the flight and carefully inserted them in the metal, link-belt segments that fell off as the cartridges entered the gun. Intense ground adjustments had to be made, with the pilot in the cockpit directing horizontal positions of the guns in conjunction with his bead-and-ring sighting on the distance to target. Our Vickers gun's rate of fire was 400 rounds per minute.

I owe my life, in a manner of speaking, to the SPAD. On one occasion, a fracas inside Hun territory found me and Smith III, number 20, low on fuel and mightily endeavoring to regain U.S. land. A Fokker DVII held me in rather tight duress but I managed to shake him by throwing my victorious but badly battered bird into a mad, full-throttle dive, at about 130 miles per hour. I made a precarious but respectable landing at an advance observation airstrip just inside our lines. That SPAD was beautiful as ever; all in one piece, as was I!

A. Raymond Brooks, opposite, stands with a SPAD XIII, its tombstones decorations marking missions flown. Capt. Brooks was 22 when he first climbed into a SPAD biplane in Vaucouleurs, France; in three months time, he had shot down six German planes. The Shooting Star insignia identifies the 22nd Aero Pursuit Squadron. In 1986, Brooks climbs into the cockpit of his SPAD XIII Smith IV *for its official dedication at the National Air and Space Museum. SPAD's cockpit, at left, before restorers began their work.*

XIII
7689
S
220HP

Burned bodies of airmen give grim testament to the pilot's nightmare—the gasoline fire. Only in 1918 did German pilots receive parachutes; the British never did. Some pilots shot themselves rather than burn, while others leaped into a free fall. An end to horror is evoked in Childe Hassam's painting of New York's Fifth Avenue, opposite, entitled "Allies Day, May, 1917," the year before the fighting actually ceased.

Longmore, and the equally absurd-looking A. D. Sparrow, which resembled an inverted Vickers Gun Bus on Adler elevator shoes.

The Germans were not immune from flights of technical fancy; Fokker built the five-winged Quintaplane in an excess of zeal and flew it only twice before hiding it. The Teutonic penchant for "schrecklichkeit," or "frightfulness" was embodied in great terror bombers with multiple engines, elaborately geared to the propellers. From this tendency came the "over and under" Zeppelin-Lindau (Dornier) Rs III and IV, the bob-nosed and nose-bobbing Siemens Forssmann, and the fork-tail Siemens-Schuckert RI, which in turn led to the grotesque Siemens-Schuckert GIII.

Yet the same firms capable of these embarrassing departures could create resounding successes that carried the portent of another age in the air. The Dornier efforts resulted in the famous line of Wal boats whose basic configuration is still being exploited today. The Zeppelin-Staaken R IV through R XV were huge and capable, and the postwar four-engine Staaken transport was a cantilever-wing, 140 mile-per-hour miracle that in quantity production might have advanced air transportation by a decade.

Flying World War I aircraft was demanding for a variety of reasons. Even at best, training was limited by modern standards, and pilots were generally supposed to be able to fly anything with wings immediately, without specialized training. Yet the aircraft were far more demanding than modern airplanes because they required constant attention. Few airplanes had any trim devices at all, so control pressures varied continuously with variations in power and airspeed. The German Gotha bombers were so tail-heavy that the pilots became fatigued and often "lost it" while trying to land. Rotary engine fighters could be deadly on takeoff and demanded careful handling at all times. With its tremendous torque and tendency for its engine to cut out, the Camel killed many student pilots during climb out.

Yet some were delightful to fly. Those who piloted the original Sopwith Pups and later replicas agree that it was a viceless, perfect aircraft. The S.E.5a and Bristol Fighter had good reputations as did the Fokker D VII.

The worst, perhaps, were those aircraft whose structural deficiencies always threatened the pilot and thus inhibited his capability to fight. The lower wing of the later Albatros fighters might twist away in a dive, while the Nieuport 28s had the habit of shedding the fabric on the upper wing. The famed Fokker Triplane and the later "Flying Razor" Fokker D VIII had quality control problems that caused wings to collapse in flight. Even the SPAD XIII, flown by most American pilots, had insoluble engine problems and a wretched glide angle. Many accidents resulted.

There was a solution that was applied in both world wars to the problem of engineering deficiency: solve it with young blood—and solve it both sides did. Yet there was another side to the coin; if the general staffs did not draw the correct conclusions from the air war, the engineering staffs did, and all over the world there sprang up clustered groups of engineers and pilots who would plot the way to the future with slide rule and test flights.

World War I was a great accelerator, forcing progress because of necessity. The next decade would see the progress analyzed, digested and improved upon. From the fire and mud and skies of World War I, a specter had emerged of the kind of air war to be fought in the future, if indeed the world were to be drawn into a second great conflagration.

NAT
U.S. MA

Boom and Bust

In 1914 in the United States, wealthy sportsmen and carnival-style aerial performers flew a few hundred aircraft. By 1919, thousands of planes were available as war surplus, with the long list of necessary support equipment, spare engines and so on. But the war had removed the sport from flying for many, and the postwar recession had taken the wherewithal to fly from others.

Despite the great advances in aviation during World War I, aircraft were still like racehorses—delicate, given to indisposition and requiring endless personal attention from a host of technically qualified people. So even when brand-new aircraft—fighters, bombers and trainers—became available, there were few takers.

Aircraft were simply too expensive except for military service—in which they were priceless. Rich individuals could and did buy and operate aircraft, but seldom for long, as

Primitive but promising, airfields of the early Twenties hosted civilian pilots of the U.S. Post Office Department and, later, employees of private contractors. Wilma Wethington's painting, left, "Threatening Weather But the Mail Must Go Through," captures the can-do tenor of the Air Mail service, as does the portrait of William "Big Bill" Hopson. He joined the service after 1920, pioneered the transcontinental route, then died in a crash in Pennsylvania before the decade was out.

these big canoes with wings and wheels and motors and propellers could cost more to run than yachts and were far more susceptible to damage and destruction. Similarly, without heavy government subsidies, commercial ventures with the available aircraft—carrying freight or passengers—faced economic disaster.

By the Twenties, though, flight had gained a life of its own, a wonderful momentum supported by the swiftly advancing science of aerodynamics, bringing profound changes in airframe design, structure, engines, equipment and piloting technique. In short, by the end of World War I, aviation had entered its early adolescence, an emotional and physical state entirely compatible with its age of 15 years. Pilots, many little older than powered flight itself, plunged into the 1920s with marvelous enthusiasm. Flight emerged a decade later, sobered and more mature, but still ready for adventure.

Perhaps a hundred factors contributed to this successful rite of passage. Some, though, were of signal importance, and each offers a special insight not only into the particular process, but also to its importance for the future. While most fliers from World War I turned with relief to other professions, a hard core remained, men for whom there could never be enough flying, men determined to create an industry, a livelihood. And as the decade passed, another group of enthusiasts emerged. They were too young to have served in the war, but they had been caught up in the romantic recollection—recounted in a hundred magazine articles—of the air war above the bloody trenchworks.

And what a ripe plum aviation was; full of records to be broken, first flights to be made, races to be won. It was a time when any individual could catapult himself (and, less frequently, herself) from the ranks of obscurity to the headlines—winning fame and fortune in a single flight. Similarly, a manufacturer housed in a garage the day before beginning a big race could, the day after, sell stock in a new multimillion-dollar company.

The only modern counterparts to aviators during this first postwar decade were the first few groups of astronauts, particularly the original seven. During the decade after World War I, any flight was considered newsworthy, with a record flight or racing triumph the clear equivalent of the world series or a championship boxing match. The war had created a need for heroes: aviation could provide them. Writer James Thurber satirized pilot-worship with his short story, "The Greatest Man in the World," about the fictitious Jackie Schmertz, who did great things in a nonstop flight around the world but was conceited and thoroughly reprehensible. The satire was well placed, for the aviator heroes of the Twenties could literally do no wrong.

Much of this, it seems, was national pride—a powerful reinforcement of the spirit of individual daring and entrepreneurial resourcefulness. Britain, for instance, bet its future on being first in the air as she had been at sea; Frenchmen made up their minds to hold all the records, to possess the largest air force, to create the international routes to knit together an "empire" less closely held than Britain's. Of course, the United States owned aviation—had invented it, produced all the best pilots. America's future was clearly painted red, white and sky blue, or so the press and public believed.

National pride manifested itself in many ways. The first and most traditional were subsidies granted to national airlines. More important were military expenditures, mostly for research and development.

The establishment of air service in Europe grew first from the need to transport dignitaries to and from Paris, where the peace negotiations were under way and, of course, for the ongoing transport of military mail and personnel. Germany began January 1919, with the Deutsche Luft-Reederei, flying passengers between Berlin and Weimar in modified A.E.G. and D.F.W. military biplanes. The work of the Treaty of Versailles had RAF Squadron 86 flying three times a day between London and Paris using Handley Page bombers and de Havilland D.H.4s. This traffic was augmented by the first regular passenger service between Paris and Brussels on

Young people of the Twenties equated flight with freedom and romance. Opposite: this woman appeared on a poster announcing "The First International Aero Congress" in Omaha, Nebraska, of November 1921. From the barnstorming days to the present, North America's Great Plainsmen and women, have welcomed fliers. In the East, especially at such Ivy League schools as Harvard, college boys banded together to share the high costs of aviation. Like their British and Continental contemporaries, they built a strong tradition of flying clubs. The rugged airmail men, though, were the ones to thrill the hearts of an entire generation of American women, as suggested above.

March 22, by Lignes Aériennes Farman, which used the Farman Goliath. The first transports were of course purely military types, but in time they came to be modified with enclosed cabins, roll-down windows, multicourse meals and wine service.

A number of factors, some not immediately obvious, fostered the spread of air routes. European countries poured money into their national airlines. America lagged behind, depending heavily on its efficient railroad network and a burgeoning system of highways. Europe also had a splendid rail system, but the airplane provided Europeans a means of transcending the routine obstacles of the Channel and the drudgery of border crossings. Wealthy people found it not only convenient but also fashionable to fly the airliners.

Alfred E. Lawson, a great, passionate eccentric, both designed the first big passenger aircraft and coined the term "airliner" which we use to this day. He and his brainchild were not only ahead of their time, but ahead of airlines, airline routes and, of course, airline passengers.

In the race for air transport supremacy, Europe's seeming advantage was later to prove to be its undoing. The very handicaps America faced in the beginning—longer distances to travel, strong competition from rail service and others—would force U.S. designers to come up with more efficient aircraft culminating with the incomparable DC-3.

National pride had a more direct expression in the development of military aviation. Almost every industrialized nation followed the same patterns. Each country developed centers to undertake the necessary research and development for military requirements. In the United States, McCook Field at Dayton, Ohio, became the center for Army Air Service development, while the Naval Aircraft Factory in Philadelphia performed a similar service for the Navy. The work of these centers was complemented by the laboratory of the National Advisory Committee for Aeronautics (NACA) in Langley, Virginia. In fact, Langley served the world; anyone could get a copy of reports on the latest airfoils, wind tunnels and more—and for as little as a dime.

In Britain, the proud and successful Royal Aircraft Establishment provided a similar nucleus of engineering expertise where bright young engineers gathered to chart the course of the future. Similar centers sprang up in Villacoublay, France, and on a lesser scale in many other lands. Perhaps their most important work was to establish requirements for the next generation of aircraft—new engines, higher speeds, greater altitude capabilities, longer range, additional safety—and then to develop the ground rules by which manufacturers were to achieve these.

Military research centers often built aircraft of their own design. Some even engaged in limited production prior to offering the designs up for bid to commercial contractors. It was a carefully balanced process, by which a number of potential manufacturers were kept alive by means of small contracts and were encouraged to create their own designs.

As it developed, U.S. commercial manufacturers became more innovative than the military centers. Consequently, the

A German national-airline photo, opposite, emphasizes wine service and dining comfort. The trimotor, at right, typifies the slow, boxy airliner of the day. By 1930, an association of flight attendants emerged, and the profession of air steward and stewardess steadily gained glamour in the eyes of the public.

centers became increasingly devoted to setting goals and to applied research, the results of which were freely dispensed to contractors. The military centers were also training grounds for engineers and entrepreneurs.

Britain had saddled itself with the "Ten Year Rule," an artificial measure that required the cabinet to peer ten years into the future to see if war was coming. If the cabinet members decided that no war was in view, military spending was kept at a low level. In the United States, it was traditional that military spending should be kept at rock-bottom levels during times of peace. France still suffered from fear of prostrate Germany, and Italy was initiating Mussolini's desire to create a second Roman empire. But everywhere, in comparison to the spending in either of the world wars, the military was kept at a very low level.

Other factors came into play to slow warbird development. For instance, as military aircraft matured, planners placed requirements for extra equipment on designers and manufacturers. Fighters began to be laden with oxygen equipment, bomb racks, radios and other gear. Airframe weight soared and performance plunged. Civil aircraft could avoid the weight penalties imposed by the military. By 1929, the issue was dramatically demonstrated at the National Air Races. Military aircraft had dominated this event for years, and this year was not expected to be any different. An Army Curtiss Hawk P-3A, specially modified and flown by an expert pilot, Lt. R.G. Breene, was the favorite, with an equally souped-up Navy Curtiss F6C-3 Hawk considered to be a strong contender. In fact the race was won by the civilian-built Travel Air Model R Mystery Ship, the first landplane to exceed 200 miles an hour with a radial engine. Walter Beech built this magnificent machine to the specifications of Herb Rawdon and Walter Burnham and it revolutionized the industry. Civil aircraft would dominate racing in the United States and abroad for nearly a decade.

Foundations were laid for enormous progress as aeronautical engineering and, perhaps more important, scientific aviation testing came into its own. The traditional wood-and-fabric methods of construction were clearly obsolete, and many methods of metal construction emerged. The first and most influential of these arose from the German experience in World War I. The Junkers and Dornier factories created first-class all metal aircraft. The postwar Junkers F 13, an all metal, low-wing aircraft, was of extraordinarily advanced design for 1919. It would have tremendous influence on aviation, and many of the 320 that were built served into the late Thirties. Dornier contributed the Delphin series of flying boats which led directly to the highly successful Wals. In Holland, Tony Fokker kept the wartime riches he had earned from Germany and retained a good portion of his factory's inventory at the end of World War I. Here, Reinhold Platz continued his basic fuselage in welded steel tubing and a wooden, cantilevered wing. Thus he started a long series of successful transports, beginning with the Fokker F.II. Angular but aerodynamically clean, the Fokkers were used all over the world, and their basic planform was imitated by a

On July 6, 1919, the British airship R 34, above, arrived at Mineola, Long Island, with 30 crew members aboard, having completed the first east-to-west crossing of the Atlantic by air. Untested before the Armistice in 1918, the Tarrant Tabor bomber fuselage, at right resembled an 11-foot-wide artillery shell. But even with six engines, this triplane dinosaur crashed in 1919 on its maiden takeoff and joined the crowded ranks of aviation's failed experiments. An ocean liner with wings—nine of them—and eight engines of 400 horsepower each, the Caproni Ca60 Transaereo of 1921, couldn't escape its watery nest, Italy's Lake Maggiore, opposite. Authorities disagree on whether it actually took off, but concur that it broke up and was never reconstructed.

number of manufacturers in both America and Europe.

Great Britain took an early and brilliant lead in all metal construction with the debut on June 3, 1921, of the Short Silver Streak with its streamlined monocoque fuselage. (Monocoque means, simply, that the skin of the aircraft provides a large part of its structural strength.) This metallic trailblazer was too radical for British thinking at the time, and the design languished.

One could note that the three revolutionary aircraft mentioned above were, in combination, the clear precursors of the future. Designers in countries all over the world would draw inspiration from them, and aircraft of the late Thirties all show elements of one or more of the three.

Postwar development of both engines and airframes had been stifled by the surplus of World War I engines. America's single greatest contribution to aviation during the Great War had been the Liberty engine, a fine 400-horsepower workhorse whose development and manufacture happened almost overnight—another example of the old American aircraft magic that had begun at Kitty Hawk. It served U.S. military aircraft well during the Twenties: this changed only when engines like the Curtiss D-12 emerged. They cost more, of course, than the "free" stored Liberties, but longer life between overhauls and easier maintenance made them less expensive in the long run. But reliability is a relative thing: as powerful and streamlined as the Curtiss engines were, they did not inspire confidence in their utility for long-distance mail and passenger flights.

There was an revolution in the making, though, stemming from the radial air-cooled engine pioneered by the American designer Charles Lawrance. More difficult to streamline than its liquid-cooled counterparts with their in-line pistons, it dispensed with the expensive and difficult to maintain coolant systems and promised greater reliability.

Charles Lawrance lacked the funds to produce his engine in quantity, and Admiral William A. Moffett of the Navy Bureau of Aeronautics virtually forced the Wright Aeronautical Corporation to absorb commercialization costs and to develop and produce the Lawrance engine. Having a heavy investment stake in its own line of water-cooled in-line engines, Wright hesitated but the Navy persisted, and to their mutual benefit.

The resulting Wright Whirlwind was a world-beater: the instant international standard. Charles Lindbergh saw the Whirlwind engine's superiority right away, and used it in the *Spirit of St. Louis* to make flight history. Rear Admiral Richard E. Byrd, Clarence D. Chamberlin and a host of others bet their lives on the Whirlwind. Frederick B. Rentschler bet his future on Whirlwind technology and left Wright in 1925 to set up his own company—Pratt & Whitney. These two pio-

neers in aircraft power plants would compete for the world market in radial air-cooled engines for the next 20 years, until jet engines revolutionized aircraft propulsion.

By the late Twenties, the modern aircraft was ready to emerge. The cantilever wing had appeared on a number of U.S. and foreign types, including the Verville Sperry R-3, the Junkers F 13, the Fokker F.II and others. Metal construction was almost at hand, although the best design methods for obtaining the necessary balance of strength and weight control had not yet been achieved.

Ethylene glycol (Prestone) cooling systems offset the radial engine's advantage. Such liquid-cooled power plants featured in-line pistons that made for lean and sharklike snouts for aircraft as contrasted with the bulbous noses required for radials. With its reduced drag and higher operating temperatures, liquid-cooling looked like a winner. A horsepower race developed between the two types. Yet it was apparent to all that the key to unlocking the full range of performance for either type lay in the creation of an adjustable-pitch propeller. With the ubiquitous fixed-pitch propeller, in use since the Wrights, the basic components of any flying machine—engine, propeller and airframe—could be optimized only for certain flight regimes. A propeller made for maximum take-off performance was not efficient for cruise, and vice versa. It was possible to switch propellers on the ground to meet certain conditions, a crutch at best. Ground-adjustable propellers helped, but clearly the next performance breakthrough depended on in-flight propeller control.

In the United States, Frank Caldwell led the way, but once again there were parallel developments all over the world. By the mid-Thirties, sophisticated controllable-pitch, constant-speed propellers were available in every country.

With the engine and propeller combination generating more horsepower, drag reduction became even more important. The cantilever wing was soon complemented by the NACA cowling designed by Fred Weick. The cowl alone increased the efficiency of such already streamlined aircraft as the Lockheed Vega by as much as 20 percent. Basically it served to cut friction of air rushing through the cooling fins of each cylinder head of a radial engine. Care was also paid to wing fillets, carefully faired (streamlined) landing gear and, in some radical instances, by fully retractable wheels.

By 1929, aviation had learned almost everything it needed to know to create the planes of the Thirties—airplanes that would for the first time begin to generate profits from their operations and, unfortunately, airplanes that would begin the process of dictating world politics.

Certain luminaries jump to mind when one thinks of the people behind the famous aircraft. Charles Lindbergh not only captured the imagination of the world but set in motion legislative and economic initiatives that influenced all of future aviation. Others were less visible than he, but also made great contributions. They, too, worked to build an institutional base for aviation.

Perhaps the most important force behind those inventions

Burning phosphorus showers a target ship, opposite, as a Martin MB-2 bomber scores a direct hit during maneuvers off the Virginia Capes in 1921. High-explosive bombs sent other vessels swiftly to the bottom. General William "Billy" Mitchell organized the demonstration of post-World War I air prowess to prove that naval vessels were vulnerable to air attack. British crowds flocked to the annual show by the Royal Air Force at Hendon. The edition above dates from 1925. Such displays of military flying skill made the United Kingdom more air-minded and gained important support for air-force modernization in the Thirties, the decade of the Great Depression.

Black racer with golden wings, a U.S. Navy Curtiss R3C-2 gleams at the Smithsonian. James "Jimmy" Doolittle stands on a pontoon of such a speedster, above. With the one at right, in mid-1920s he took the Schneider Trophy, an international air race with strong military implications. The brass radiators of the floatplane's water-cooled, V-12 engine were built into the wings to reduce aerodynamic drag. Unlike the era's more familiar air-cooled (radial) engines, in which the wind passed over a circle of cylinder heads to keep temperatures safe, the Curtiss engine was designed to be long and narrow for streamlining. Doolittle's 1925 record of 245 miles per hour was nearly doubled by 1934, largely through intensive European development of liquid-cooled engines and monoplane, all metal airframes. These advantages found their way into the generation of fighter planes with which Britain, Germany and others combatants entered World War II.

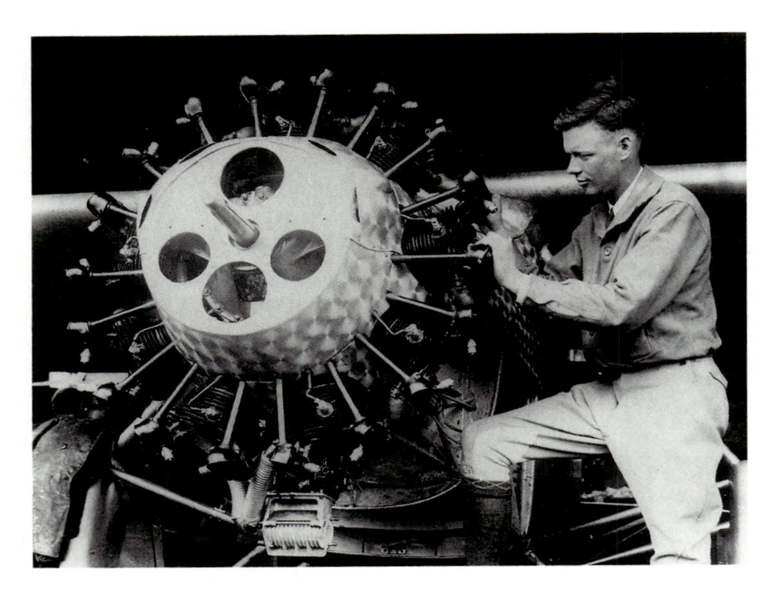

Though outwardly easygoing, Charles Augustus Lindbergh was totally focused in his quest for technical excellence. This trait led him to the Ryan company of San Diego for the design and construction of his Spirit of St. Louis. *Without the new Wright J-5 Whirlwind engine, above, Lindbergh's flight in 1927 might well have ended in disaster. The 25-year-old airmail pilot recognized the Whirlwind as the first really reliable aircraft engine.*

of the Twenties was the creation of the National Advisory Committee for Aeronautics (NACA), formed in 1915, the ancestor of NASA (National Aeronautics and Space Administration). NACA began, curiously enough, in the "old tin shed" that subsequently became the National Air Museum of the Smithsonian Institution. In 1919, Charles D. Walcott, the fourth Secretary of the Smithsonian, who had been Chairman of the organization's Executive Committee since its inception, was designated Chairman of NACA. In 1916, NACA recommended the establishment of an airmail service. Otto Praeger acted upon the suggestion. He aggressively orchestrated the Post Office, Congress, the NACA and the U.S. Army Air Service to establish an airmail service that he believed would be fundamental in wedding the U.S. government to the advancement of aviation.

After a shaky start on May 15, 1918, (one of the pilots got lost immediately after takeoff) the Post Office had by September 8, 1920, established a transcontinental route from Hazelhurst Field, New York, to San Francisco. Even five years earlier, such a great leap as a coast-to-coast airmail service would have seemed impossible.

The natural tendency toward "privatization" of the American business community, together with a number of associated factors, resulted in the Contract Air Mail Act (the Kelly Act) of February 2, 1925. It provided for the airmail service to be transferred to private operators on a competitive bidding basis. It marked the start of a viable aircraft industry in the United States.

The military services also were reassessing their future roles in aviation. Chief of Naval Operations, Admiral E. W. Eberle, took the lead for his service. The famed Eberle Board seemed to have a traditional battleship orientation, but actually laid the groundwork for the Navy's great aerial armadas of World War II. To join the tiny *Langley,* the Navy ordered the completion of the *Lexington* and the *Saratoga* as aircraft carriers and initiated a program of further aircraft carrier construction and aircraft building. The Army's Lassiter Board conducted a similar introspective investigation, recommending the creation of an independent group of bombers and fighters for employment by the commander of the Army General Headquarters. And in the Congress, anticipating history by 20 years, the Lampert Committee recommended the establishment of a unified air force, independent of the Army and the Navy, and a Department of Defense.

Another board, of broader constituency, was the Presi-

Spirit of St. Louis, *a flying gasoline tank with a superb power plant, gave early proof that aerial machines could match the dreams of daring aviators. After the nonstop flight to Paris and visits to other countries, the* Spirit, *below, was brought back home to a place of high honor in the Smithsonian collections; it is now displayed at the National Air and Space Museum.*

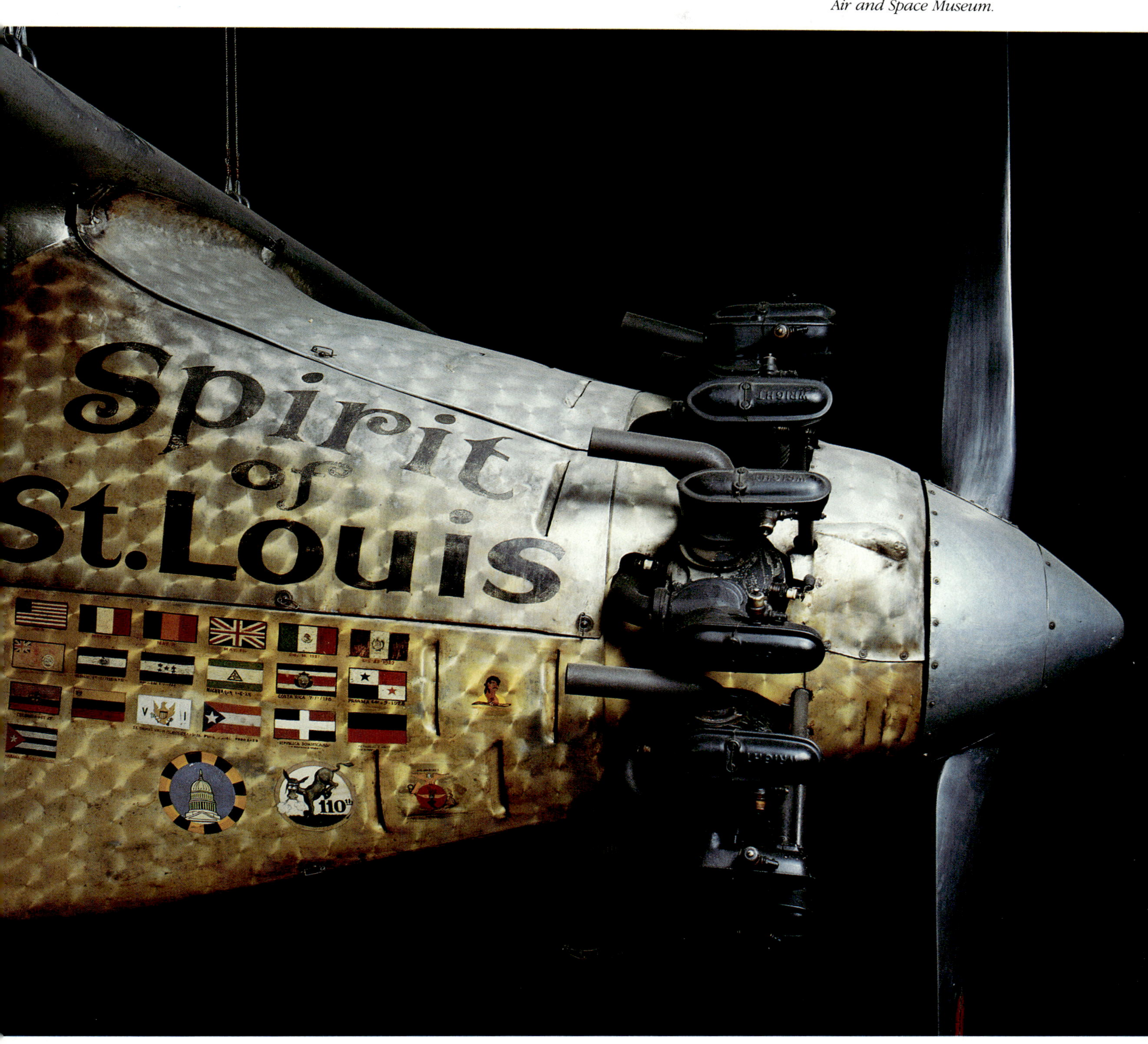

Embraced by the world, as well as by the United States ambassador to France, Myron T. Herrick, at right, Lindbergh was lionized abroad. His triumphant return to New York helped set the style for all ticker-tape parades to follow.

dent's Aircraft Board, known more generally as the Morrow Board after its chairman, Dwight Morrow. Proponents of independent air power took heed. This body rejected the idea of an independent air arm, instead reducing its size, but did change the name of the Army Air Service to the Army Air Corps to reflect its broader functions. The board also called for the formation of separate civil national authority for aviation under the Department of Commerce, to be called the Bureau of Aeronautics.

The Air Commerce Act, which President Calvin Coolidge signed on May 20, 1926, laid the foundation for subsequent civil airline history. A bitter period followed in which air mail contracts became political footballs, with contractors charged with various irregularities. Such wrangling resulted in the infamous 1934 decision of President Franklin Roosevelt to have the Army Air Corps carry the mail. It was a debacle: about a dozen Army pilots, poorly trained for bad weather and night flying, were killed in a brief period. This tragedy, highlighting the need for further regulation, contributed to the Civil Aeronautics Act of 1938.

At the same time individuals like Praeger, Kelly, Eberle and Morrow were creating the framework for the orderly growth of aviation, other individuals were investing their lives and fortunes in the establishment of aircraft manufacturing plants, airlines and every sort of support activity. There has been a curious reversal in the trends of the establishment of the companies over the years. In the beginning, there were literally hundreds of aircraft manufacturers, who came and went with regularity, and of whom only a very few survive today—Boeing, Lockheed, McDonnell-Douglas and perhaps half a dozen others. In contrast, relatively few airlines emerged and most of these were absorbed by the force of competition and legislation into the "majors," companies such as United Air Lines, American Airlines, Pan American Airlines and TWA (Trans World Airlines)—until, of course, deregulation of the airlines during the early Eighties. By 1986, moves toward merger occurred, as with the amalgamation of Republic Airlines and Northwest Orient.

Times change, and legislation changes, but the turbulent forces of the marketplace are always constant. In terms of the human element, men such as Eddie Rickenbacker and Juan Trippe created the airlines and men such as Reuben Fleet and Donald Douglas built up the aircraft manufacturing industry. They differed from the ace pilots of the day only in the nature of the risks they took. In many respects it took as much courage to collect money from friends and the public to

Literally overnight, the unknown Lindbergh became the world's darling—as soon proclaimed in song, at top. An elaborately illuminated check, above, represents his prize money for winning the Raymond Orteig challenge to fly nonstop between New York and Paris, in either direction. Though Lindbergh was the 79th person to cross the Atlantic by air, his was the first solo, nonstop leap between the North American and the European mainlands.

Boxy passenger planes and giant airships, below, soon became obsolete, though they persisted into the Thirties.

invest in an industry that had a reputation for absorbing those funds endlessly without return as it did to fly the "hot ships" of the day. The goals were the same; all of them wanted to make money, of course, but that was incidental to the creation of an aviation industry for America.

In noncommunist countries the same system was followed. Individual air exploits by "heroic" pilots were matched by far-sighted legislative efforts and entrepreneurial gambles.

In some instances, strangers to the scene supplied the necessary catalyst to keep things going. In the United States, the Daniel Guggenheim Foundation for the Promotion of Aeronautics was fundamental in the development of blind-flying instrumentation, the establishment of the Safe Aircraft Competition and perhaps, most importantly, in the founding of major centers of aeronautical engineering at seven major U.S. universities.

Sometimes philanthropy was related more to national pride than to devotion to aeronautics. In Britain in 1930, the Labour Government had concluded that it could no longer fund competition in the Schneider Trophy races, even though Britain needed only one more victory to retire the trophy permanently, having won it in 1927 and 1929. There was a furor, and Lady Lucy Houston stepped forward with an offer of £100,000 for development and the statement, "I am utterly weary of the lie-down-and-kick-me attitude of the Socialist Government. . . . I live for England and I want to see England always on top." And on top she saw it, for her £100,000 not only procured the 1931 Schneider Trophy-winning Supermarine S.6B, but set in motion the design thinking that would result in the Spitfire fighter with its wizard of an engine, the Merlin, produced by Rolls-Royce. This engineering package and its progeny would be instrumental in winning World War II.

The most obvious human element of this turbulent decade of progress was that provided by the many heroic pilots who almost every day seemed to set new records for flying farther, faster, higher and longer. Yet the pilots increasingly depended on science and engineering. All around the world, the eagerness and desire of individuals were advancing aviation to levels not yet perceived. A Golden Age was on the way, the last ten years before World War II, the glorious decade when aviation came into its young maturity.

The Travel Air Model R Mystery Ship, at left, was a harbinger of aircraft to come. At the National Air Races of 1929, held at Cleveland, Ohio, it vanquished even the military entries, winning the free-for-all speed competition through several innovations, including a NACA cowling around its radial engine to cut air resistance. It was the first ship with an air-cooled, radial engine to surpass 200 miles-per-hour.

NES
EASTERN
Air Lines
DOUGLAS
DC 3
12
NC11Y
TWA

The Golden Age

Despite the Great Depression, aircraft research and development advanced rapidly during the Thirties. The moving force was national pride and the desire of nations to be prepared for war.

All metal airplanes with single wings and advanced engines came into production during this decade, but such progress during times of scarcity was not a blossom without roots. Without the industrial and scientific momentum of the Twenties—the era of surging profits, investment and philanthropy—the aviation industry would have languished along with so many others.

Perhaps the Thirties can only be called a Golden Age in retrospect: even some of the major manufacturers scraped by on small government contracts. Yet during these generally impecunious times, designers and manufacturers were beginning to gain strength from the new scholarly discipline of aeronautical engineering. The old art of cut-and-try, practiced by the pioneers of flight since the Wrights, was as good as dead, though it had a few graceful swan songs.

As suggested in the previous chapter, the British just squeaked by. Lady Houston, who saved the day for the Schneider Trophy race of 1931, saw the British entry win with a world speed record of 379.05 miles an hour. But there was no time to rest on laurels. In Italy, for instance, a 1934 Macchi Castoldi M.C. 72 racer set a new record of 440.68 miles an

Direct ancestor of a famous family of Douglas passenger planes, below at left, was the Northrop Alpha of 1930. The Douglas DC-3 at top bears the family resemblance with its design for engine cowling, wing and tail. Both aircraft feature an all metal structure of the type originally developed by John Northrop. In half a century, a worldwide network of airlines, airports and routes grew from such humble beginnings, as indicated by the crop of baggage claim checks, below.

J P WIERTZ
IN 2 DAYS TO NORTH AMERICA!
DEUTSCHE ZEPPELIN-REEDEREI

Nearly two-thirds as long as the Empire State Building is high, the Hindenburg *(seen on a travel poster opposite) last showed itself to Manhattan at 3:00 p.m. on May 6, 1937. At 7:20 that evening, over Lakehurst, New Jersey, this greatest of the dirigible airships burst into flame.*

hour. And while the United States placed somewhat less emphasis on speed, America was on the move in ways that would place it at the forefront of world aviation by 1945 and the end of World War II. In fact, during the year after Pearl Harbor—1942—the Americans had flown all the prototypes of all the aircraft that they would mass produce for the war effort. Several types had been either on the drawing boards or in the air before Germany's blitz of Poland in 1939.

Aeronautical engineering in the United States began in 1911 with the graduation of the first American aeronautical engineer, Grover Loening, from Columbia University. By 1930 a number of engineers had added academic honors to their aviator's wings. Best known was James Doolittle, whose Doctor of Science in Aeronautics degree from Massachusetts Institute of Technology prepared him for a lifetime of scientific achievement. Doolittle's piloting feats were so remarkable that they have always overshadowed his scientific accomplishments. He won races, set records, led the 1942 carrier-launched raid on Tokyo and commanded major air units during World War II. Of a more scientific nature, he helped to pioneer instrument flight and worked to develop 100-octane aviation fuel.

The drama of change was so enthralling because so much improvement still needed to be made. Aeronautical engineering was young enough that intuitive breakthroughs were still possible. Twenty years later, each minor improvement rested upon mathematical calculations and exhaustive testing. The early Thirties were a time of transition that saw the intuitive designer with his hunches and gut feelings gradually supplanted and surpassed by the academically trained engineer.

The potent and dangerous Gee Bee speedsters that dominated races of the Thirties were a combination of Zantford "Granny" Granville's intuition and the engineering backup of Robert Hall and later Howell Miller. Granville thought that an aircraft of teardrop shape represented the ultimate in streamlining, a shape whose maximum diameter lay 34 percent aft on the fuselage. He designed his airplanes accordingly, tiny short-winged brutes with giant engines that embodied power, speed and danger. Miller, along with professor Alexander Klemin, validated his ideas through wind tunnel tests at New York University. And Granville's intuition proved to be correct for the racing speeds he intended for his airplanes. Doolittle and other Gee Bee pilots achieved the speeds that tests in the wind tunnel had indicated were possible.

Germany established eastern routes, linking Berlin with Soviet cities. As part of a secret accord of the Twenties, the Russian city of Lipetsk played host to a reborn German Air Force. Such warlike training had been forbidden by the Treaty of Versailles of 1919.

Overleaf: Whatever their vintage, classic sports planes possess timeless appeal. For the picnicking couple aviation's first Golden Age began early, with their Travel Air biplane of the mid-Twenties. Cherished and polished, many antique ships fly even today.

MAIDEN WICHITA

The products of designers such as the team of Granville and Miller had a glorious flavor, a zest and flair that dominated the minds of racing enthusiasts. On the other hand, Jimmy Wedell created the best racers of the Thirties. They won more often, and with greater safety than the Gee Bees did, yet the look of them left audiences cold. Wedell's circumstances were different than Granville's, for he had the considerable funds of Harry Williams at his disposal, but his designs were strictly his own, sometimes marked on the hangar floor in chalk rather than in blueprints. At least, that's what legend says. His success seems to have been a case of the triumph of pure intuition.

With the death of Zantford Granville, Howell Miller enormously improved the Gee Bee formula in his *Time Flies,* built for Frank Hawks; it was one of the most aerodynamically advanced aircraft of the time. Howard Hughes' H-1 racer was, and still is, the ultimate expression of gut-feel design supported by aeronautical engineering. With engineering backup from Richard Palmer and Glen Odekirk, Hughes distilled into a single airplane all the most advanced techniques of his day. He and his team created what many believe to be the most beautiful piston-engine aircraft. With Hughes' money and his fanatical devotion to perfection, they built an aircraft that set both the transcontinental record in 1937 and speed record for a measured course in 1935.

America's style in air feats so often favored the likeable loner; though with Howard Hughes it was a case of the high and the mighty. Europe had a little something for everybody; England's Kings Cup, for instance. Almost anybody could enter this great gregarious steeplechase in the sky. It turned aerial handicapping into a highly developed art form. And just like European soccer today, every race had its nationalistic overtones. The Coupe Deutsch races called forth the best efforts of France's factories while the Schneider Trophy races fanned fierce competition among the French, Italians, British and Americans. The individual racers clearly were developmental vehicles for military engines and aircraft.

The wild and earthy permutations of these early engineers set a style for the Thirties, but it was short-lived. The fast racers at Cleveland and elsewhere had in effect consumed the fat of aeronautic development during the Twenties. Builders simply created tiny specialized airframes into which they crammed big engines with controllable-pitch propellers and NACA cowlings. By the mid-Thirties the retractable landing gear had just begun to appear.

Between 1930 and 1937, the fastest of these civilian air-

In artwork by John Paul Jones, Charles and Anne Morrow Lindbergh prepare their plane for a survey flight around the Atlantic. In the early Thirties, these newlyweds made two notable aerial voyages in the craft, in both cases helping to develop transoceanic routes—called Lindbergh Trails. In 1931 they flew across the arctic to Japan and China; in 1933 they traveled in Europe, Russia, Africa and South America.

planes exceeded military aircraft in speed—but in speed alone. During the same time, factory design and engineering teams were creating airplanes that would dominate the Forties, clearly surpassing the amateur efforts that had made aviation so romantic. Aeronautical engineering not only enabled the industry to make breakthroughs, but to make breakthroughs a routine and expected part of the business. Aircraft developed for either airline or military use represented quantum leaps, markedly exceeding earlier performance. Soon the fully engineered machines were achieving vastly higher levels of speed, range and altitude capability. Not until the mid-Seventies would the efforts of amateur builders regain importance.

Companies soon adapted to the new environment or suffered for it. Grumman, Douglas, Martin, Boeing and Consolidated and a few others thrived. As the demands for performance became more rigorous, they simply built better aircraft. Some, like giant Curtiss-Wright, seemed more interested in finances than engineering, and lost ground in both areas. Still others, including Keystone, Berliner-Joyce, Great Lakes and Thomas Morse, simply could not cope; their losses mounted and they either shut their doors or found themselves absorbed by other firms.

Economic and political realities dictated a change in the style of the aviation business, and gave the industry and its leadership very close ties to the military. This transformation turned upon three factors. The first was a legacy from World War I. The surplus inventory of Liberty and OX-5 engines and the Jenny and Standard airframes had inhibited industry, depressed prices and kept performance expectations low. Then came the profit motive. Aircraft were just beginning to offer the promise of legitimate, unsubsidized profit. Speed and carrying capacity improved, and instrument flight began to develop. Third, World War II became inevitable. Politicians demanded and received higher and higher performance from both civilian and military craft. It had long been clear that if war came, it would begin in the air and very likely be ended by aerial action.

Now our world has become inured to the prospect of instantaneous global nuclear devastation. Postwar generations find it hard to imagine the dread that possessed both the public and politicians during the Thirties. It was as if World War I air warfare had been distilled so that the reality evaporated and only the propaganda remained. Potential allies seemed hooked on fear and reinforced each other's war jitters. In fact, the British and the French politicians freely

The Lindberghs don electrically heated flight suits for far-north navigation. The poster above gives a hint of the romance and excitement that air travel evoked in the pre-World War II era.

admitted—trumpeted, in the words of the British Prime Minister Stanley Baldwin, "the bomber will always get through." Hysteria was the order of the day.

Both sides anticipated the complete obliteration of London and Paris within a few hours of the outbreak of war. The Germans and Italians played on the fears of the foe. Their puppet presses boasted about the damage their air arms would deliver. The most famous quote in this war of words is attributed to Hermann Göring, chief of the reborn German Air Force, who promised that Germans could "call me Meier" if one enemy bomber penetrated Reich territory even as far as the Ruhr.

Objective evidence clearly indicated that no nation in the world could mount a decisive bombing attack, or even an effective one. Discounting all the fascist braggadocio, trusting the experience of the first war, adding up postwar practice results and factual estimates of foreign air fleets would have revealed this truth. In 1936, Germany could mount only a one-way mission against London.

Bigger and better seaplanes were required to fly the routes set by the Lindberghs. Left, Igor Sikorsky's S-40 was an important early step in the direction of the sleek Clippers that appeared toward decade's end. At the controls of the S-40, in a snapshot by Igor Sikorsky, Charles Lindbergh shows his delight with the feel of the new flying boat.

Two hundred obsolescent aircraft made up Hitler's operational air fleet. The woefully inadequate Dornier Do 23 and the bomber version of the Junkers Ju 52 could haul 2,000 pounds of bombs or fly a range of 800 miles—but they could not do both on the same mission. In desperation they might have loaded 1,000 pounds of bombs and enough fuel for the trip out from bases on the German mainland, but the effect would have been slight. London would scarcely have been marred even if all 200 bombers had reached this target. At the time, the English capital, the biggest target on the globe, still resembled a big village green adorned with scattered clumps of city, monument and palace, and sprawling factory and wharf areas. Yet, when war finally came, London was able to withstand a far greater weight of attack over and over again. The Blitz proved the mettle of Londoners and, indeed, the Berliners themselves would show how tough a civilian population could be under bombardment.

What is more, until the latter days of World War II, bombers could not hit specific targets. When raids commenced during World War II, intelligence officers could not believe the poor results. At the onset of hostilities no one had any idea of the proficiency required at all levels—from air chief marshal to privates first class—to render air power effective.

At the time, Paris was perhaps a little more vulnerable than London, but not much. Yet British and French leaders trembled in their boots, convinced that the Germans—and to a lesser degree the Italians—could deliver devastating and war-winning blows from the air. Leaders of the soon-to-be-Allied air forces saw the threat analysis as a driving factor in their own programs of preparedness. They were not about to belittle the "Hun" and his friends.

The Germans fully exploited the fears of the Allies. For instance, the Luftwaffe leaders led the French Chief of Air Staff General Joseph Vuillemin on a dazzling tour of their fields. They had picked up on a Mussolini trick by flying all the Reich's aircraft to one airfield to exaggerate their numbers. The ploy worked, and Vuillemin returned home to advise his premier that in the event of war the French Air Force would be wiped out in one week. It was all a policy of deception, orchestrated from the highest levels of government. Thus we behold the greatest air leader of the period, the man who achieved the most with the least—neither Douhet nor Mitchell nor even Trenchard—but instead the clubfooted nonpilot Joseph Goebbels—master of the Big Lie as Minister of Propaganda under Hitler.

With the help of Goebbels, Hitler scored early successes—from the bluff of the reoccupation of the Rhineland, the bloodless bullying acquisitions of Austria, Czechoslovakia and Memel, to the months of the Phony War in France after Poland was defeated. All were ultimately more than repaid, because the German victories forced the Allied powers to attempt to rearm at a pace they would have considered impossible before.

Never has a poorly armed, ill-trained, imperfectly disposed and semiobsolete force ever intimidated the world as the Luftwaffe did. Hermann Göring, chief of the new German Air Force, rattled his paper sabre for Adolf Hitler as no one ever had and, with it, the *Führer* and his minions marched the world toward a second great conflict.

In fairness, the prowess of the press in shaping public opinion should be noted: their role in the Thirties foreshadowed media influence in the Vietnam era. For instance, there were many bitter conflicts between the two global wars:

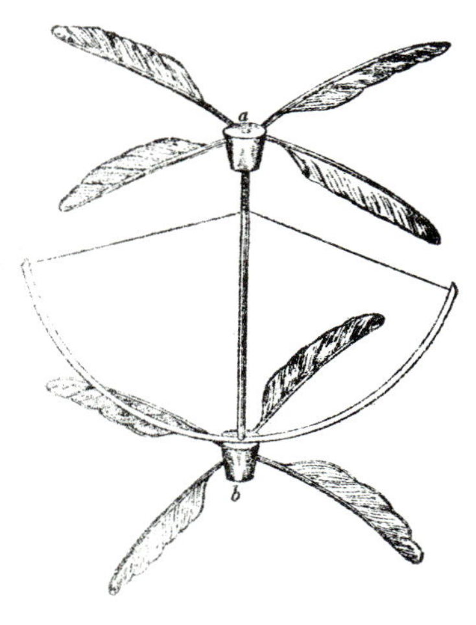

Opposite, with Sikorsky at the controls, his VS-300 took to the sky on September 14, 1939. In 1909, in his native Russia, at left, this vertical flight machine made a skip and a hop. The first craft in the evolution of the modern helicopter was perhaps a toy, as above. In Paris in 1784, experimenters Launoy and Bienvenu made such a craft with a bow-and-string motor and feather propellers.

the Gran Chaco in South America; the disgraceful, embarrassing Italian punishment of Ethiopia; the Sino-Japanese nightmare with its ancillary Russo-Japanese conflict in Manchuria; and Spain's bitter Civil War. In each, air power appeared in traditional and largely ineffective ways. Yet, certain key dramas received enormous attention. These included the bombing of the U.S.S. *Panay,* the terrorist bombing of Chinese cities (who could forget that heart-wrenching vignette of a bruised and torn Chinese baby sitting in the rubble crying?) and of course Guernica. Each such bloodletting permitted the press to reinforce fascist claims of obliterating power.

We find, though, a nugget of truth in these outlandish assertions: the progress of air power held the promise of superweapons. In this, Hitler's emphasis was correct, though his timing was tragically flawed, his resources inadequate.

In many ways, 1934 and 1935 were the watershed years during this period of golden days. At drafting tables all over the world engineers sought different solutions to different problems, hoping to achieve a new quality in air power. No longer was it sufficient to create a design with a competitive performance. The aircraft also had to be amenable to mass production and easy maintenance and repair, especially under difficult conditions in the field.

During the mid-Thirties, important designs sprang from the drafting tables of creative engineers all over the world; the flood of genius was not confined to the great powers, but it naturally flourished there. In Germany during 1934, Wilhelm "Willy" Messerschmitt and Walter Rethel first laid down the angular lines of the Messerschmitt Bf 109 fighter, hoping to cram a "huge" engine of nearly 700 horsepower into the smallest airframe possible. They did not dream that highly modified descendants with 1,800 horsepower engines would still be in production 11 years later. More than 33,000 would be built, and production would continue in Spain and Czechoslovakia long after the Luftwaffe had folded its wings. With its low cantilever wing, retractable landing gear, all metal construction and enclosed cockpit, the Messerschmitt Bf 109 would define the piston-engine fighter for World War II.

In Britain, two epic designs were created, both essentially in advance of official specifications. Sidney Camm broke his design team away from a long series of beautiful biplanes to fashion the Hawker Hurricane which, with the exception of its tubular metal construction and fabric covering, had all the Messerschmitt advances. And at the Supermarine Aviation Works, R. J. Mitchell took the same 1,000 horesepower Rolls-Royce Merlin engine that the Hurricane used and built around it the supremely beautiful eight-gun Spitfire.

In the United States, Curtiss-Wright had been the leading manufacturer of fighter aircraft for many years but had seen its position eroded by designs from Boeing. In 1934 Curtiss

The Douglas Ship of Destiny

Within two weeks of its commercial introduction in 1934, the 14 passenger DC-2 broke four speed records. Even so, the DC-2 was only a warm-up for its successor—the 21-seat DC-3—an aircraft that would revolutionize commercial transport.

On March 31, 1931, a tragedy rocked the nation: Knute Rockne, Notre Dame's popular football coach, was killed when the Transcontinental & World Airlines (TWA) Fokker F XA transport carrying him and seven others crashed in Bazaar, Kansas. Shortly afterward, the Aeronautics Branch of the Department of Commerce required expensive inspection of all Fokker F XA wings. The plywood used in the wings had delaminated, or split in layers.

TWA elected to replace its Fokkers with new airliners and sent out a list of specifications for all metal aircraft to several manufacturers. Douglas Aircraft Company responded with the DC-1, prototype of perhaps the most successful airplane of all time, the DC-3—the first aircraft that made it possible for airlines to operate profitably without airmail subsidies.

The DC-3 represented a quantum leap from the generation of previous planes such as the Ford Tri-motor and Fokker: five and a half years after its inaugural service in 1936, the DC-3 provided 80 percent of the domestic scheduled service. Its near monopoly of the prewar airliner market was marked by a strong safety record; during the peak of its domination of the market in 1939–1940, the DC-3 enjoyed a perfect safety record.

While the tremendous success of the DC-3 is the result of the work of thousands of people, one name stands out above the rest: Arthur E. Raymond, chief engineer of the project at Douglas. He recalls his trip with Douglas's General Manager Harry Wetzel from California to meet TWA president Jack Frye in New York to discuss the new design: "We went by rail, for I needed time and concentration to develop my performance figures—and we wanted to get there," a sign of the problems with safety and convenience that plagued commercial air travel at the time.

In addition to his duties at Douglas, Raymond taught aeronautical engineering at the California Institute of Technology. The cooperation between the industry and its hands-on expertise in construction and design and the university with its academic work in fluid mechanics, mathematics and other theoretical studies resulted in a fruitful, symbiotic relationship.

The DC-3 used many of the technological advances of the day: metal skin monocoque construction replacing steel tubing and cloth or wood and cloth; internally braced monoplane design instead of external bracing; the new NACA cowling that reduced the drag of the engine by almost completely enclosing it; the addition of wing flaps that made it possible to increase cruising speed without changing the landing speed; two-pitch and later variable pitch metal propellers.

Between 12,000 and 13,000 DC-3s, civil and military, were produced. Even after the DC-3 was overshadowed by later aircraft, feeder airlines all over the world still relied on the DC-3, and hundreds still fly today, more than four decades after production was stopped—a great testament to the reliability of this enduring aircraft.

hired Don Berlin to rectify the situation, and he did so with the Hawk 75, an aircraft that led first to the radial-engine P-36 and later to the famous P-40 series of fighters with which America began the war. The P-40, like the Hurricane, lagged behind the Spitfire and the Messerschmitt in performance, but pilots appreciated this rugged airplane, which served on every front.

Similar efforts occurred elsewhere to develop new generations of aircraft, although not all achieved the performance of the first four mentioned. In Russia, the barrel-shaped Polikarpov I-16 was already in production, and in the early Thirties was probably the most advanced fighter in the world. Along the Manchurian border it would meet in combat the Japanese Mitsubishi A5M "Claude," designed in 1934 and the ancestor of the famous Zero. Poland's series of attractive gull-winged fighters led to the PZL P.11, which set a speed record in 1934. Italy with its Fiat C.R.32 and Czechoslovakia with its Avia B 534 would perpetuate the biplane formula.

In 1934 bomber designs also flourished. The Germans pulled their old ploy—"we're really building a transport"—with the Heinkel He 111 and Dornier Do 17. Nothing could hide their intentions for the notorious Junkers Ju 87 Stuka. The Italians turned to Savoia-Marchetti for perhaps their best aircraft of the war, the S.M.79 bomber. A handsome trimotor of antiquated construction but excellent performance, this craft is regarded by Italian aviation buffs as the Spitfire is by their British counterparts.

During these vintage times, the Allies produced a mixed bag of bombers. In patriotic zeal, Great Britain's Lord Rothermere demanded a high-speed passenger transport, the Bristol Type 142; he named it "Britain First" and promptly donated the design, which became the disappointing Bristol Blenheim bomber. And in Seattle, Washington State, Boeing's talented team of engineers were lofting plans for what would become the B-17 Flying Fortress.

Commercial aviation made similar progress. Boeing had done its B-17 homework on a series of designs that had culminated in the Boeing 247, a revolutionary transport that appeared in February 1933. Suddenly every other passenger transport in the world was obsolete. The 247 included all metal construction, a cantilever wing, retractable landing gear and neatly cowled engines. And yet Boeing, on advice of its airline pilot experts, had made the plane too small; it could carry only ten passengers. United Airlines bought the first 60 Boeing 247s anyway. Then TWA requested a response from Douglas. First came the DC-1 (of which only one was built), then the 14-seat DC-2 and finally the 21-passenger DC-3.

The new Douglas airliner was everything the Boeing was and more. The 247s were quickly phased out of production and relegated to lesser routes. In peacetime, the DC-3 was the first transport aircraft capable of making a profit without a subsidy; when the war began a few years later, the DC-3 became the C-47 and altogether nearly 13,000 civil and military versions were built. The DC-3 was the basis for the growth of airlines all over the world. Now systems could be

Air France's Dewoitine D-338 trimotor, at top, reduced flying time between Paris and London to 90 minutes in 1936. Air Orient, affiliated with Air France, helped hold together France's prewar global empire.

Overleaf: Setting the transcontinental speed record in 1937 with his H-1, multimillionaire industrialist and moviemaker Howard Hughes averaged 327 miles per hour between Los Angeles and Newark, New Jersey.

R
258Y

In 1935, Wiley Post, below, pioneered in the use of high-altitude pressure suits and pushed his commercially manufactured Lockheed Vega, Winnie Mae, *into the jet stream at 40,000 feet, improving both distance and speed.*

American humorist Will Rogers championed commercial flight; he appears with an airline stewardess, above. In 1935, Post and Rogers teamed up on a polar flight; they lost their lives in a crash at Point Barrow, Alaska.

established that offered regularity, speed and safety to the passengers—and profit to the owners.

The DC-3 inspired imitation everywhere. Russia and Japan simply cloned the DC-3; companies in France (Bloch 220), Russia (ANT-35) and Italy (Fiat G.18V) copied its lines so closely that one would have to squint to see the difference. Yet nowhere did anyone capture the performance, the durability and especially the élan of the American product. It was, and remains, simply superb.

Boeing matched its investment in the B-17 with a gamble on a new transport, one far more ambitious than the DC-3. It was the model 307, the Stratoliner, the first civil airliner to feature a pressurized cabin. The four-engine aircraft was considered so sophisticated that an extra crew member, a flight engineer, was required. The market in 1939 was not quite ready, so only ten were built. But they clearly pointed the way to the future.

One of the more advanced transports of the Thirties, the Focke Wulfe Fw 200 Condor, accommodated 26 passengers and broke records with its round-trip flight to New York in 1938. In those times of tension the flight's ominous overtones were not ignored. During wartime, the Luftwaffe pressed the Condor into service as a bomber and a reconnaissance plane and teamed it with submarine wolf packs to operate against Allied convoys. It remained in use as a transport until the end of the war.

The pace of private aviation lagged behind that of the military and the airlines. Traditional methods of construction—metal or wooden wings, fabric covering and fuselage of welded steel tubes—were adequate, especially with enclosed cabins. Fixed, "panted," landing gear, as that on the lovely cabin Wacos, also proved to be sufficient. The lines were magnificent, and the Stinson SR-10 "Gull Wing" was easily the equivalent of the later Beechcraft Bonanza, while the beautiful biplane Beech "Staggerwing" corresponded in prestige to today's Learjet. Yet even in this arena the future was forecast. Spartan created the masterpiece Executive, a fully modern all metal and low-wing cantilever aircraft with retractable landing gear and controllable-pitch propeller. It was destined to become an object of desire of all pilots for all time. Beech also presented its "twin Beech" Model 18, the first true executive aircraft. It had all the features of the DC-3 in smaller form, and once its rather tricky ground-handling characteristics were mastered, was a delightful aircraft to fly.

The coming of the war expanded the market for the Spartan only slightly, but the Beech Model 18 became militarized in a whole variety of guises, particularly as the C-45 "Bugsmasher." It would be in production longer than any other civil aircraft—32 years—a record that would be exceeded only by the postwar Bonanza.

There were similar, brilliantly designed foreign aircraft. In Britain, famous novelist Neville Shute Norway helped produce the Airspeed Courier designed by A. H. Tiltman—it was the first British production civil aircraft to have retractable landing gear. Avro built a series of Waco-like biplanes while de Havilland produced a wide variety of aircraft, all of

Roscoe Turner's enduring claims to fame were his three wins of the Thompson Trophy races, the most collected by a single aviator. After tasting the lion's share of victory, Turner retired to more sedate diversions—going into the flying business in Indianapolis, which he felt was the only city where he wouldn't be cheated. On the Time cover of October 29, 1934, the famed newsmagazine declared, "His merchandise is speed."

Showing just how far a superb pilot can fly on a sunny smile and a custom aircraft—the Meteor*—Roscoe Turner took America by storm, and a decade after the Barnstorming era. Flying in the cockpit (at least until he outgrew it) Turner's mascot, Gilmore, was a valued part of the act. But this one-man, one-cat flying circus would never have gotten off the ground without Turner's solid skills, demonstrated in tough competition.*

Left, with rubber band-powered model, designer Sidney Camm added immeasurably to Britain's survivability through his Hawker Hurricane fighter. Basically, it was his Hawker Hart (of the sort seen in the Himalayas, below) with the top wing removed. With the Rolls-Royce engine developed from Supermarine's Schneider winner of 1931, Camm's tough airplane worked alongside the Spitfire to turn back Hermann Göring's Luftwaffe in the Battle of Britain. (See Hurricane and Spitfire flying together on pages 4 and 5 of this volume.) Camm also worked on the Harrier line—vertical takeoff jets that were effective in the Falklands War of 1982 and which serve with the United States Marine Corps as the AV-8A and B.

Speed merchants on pontoons, entrants to the Schneider Cup broke world speed records time after time. Britain, though, came close to losing the biggest race of all—the one for national survival. Lady Lucy Houston, below, came to the aid of her country, putting up money for the last British entry. The winner, Supermarine's marvelous S.6B—with its Rolls-Royce engine—led to the development of the Rolls-Royce powered Supermarine Spitfire eight-gun fighter, and just in time to fend off German invaders in 1940. If the British fighter pilot ever had a patron saint, it was Lady Houston.

robust construction and delicate line. One of the most intriguing—although built only in small numbers—was the Heston Phoenix, a rakish monoplane with an "undercart" that retracted into almost sesquiplane-size fairings. And for the pure enthusiast there was the sleek line of small, fast Percival Gulls.

France deployed a similar array of aircraft, including a tiny, fast Caudron clearly patterned after the Coupe Deutsch racers. More than 500 examples of the four-seater Simoun were built. In Italy, civil aircraft appeared in small numbers but in great variety, and with typically Italian beauty.

But, not surprisingly, it was in Germany that the most advanced civilian aircraft was created, the Messerschmitt Bf 108 Taifun. Precursor of the 109 fighter, it set speed records before the war, served on every front, was manufactured by the French in postwar years, and now appears in almost every movie requiring a simulated Nazi fighter plane. It was fast, reliable and beautiful and is still much desired by antique warplane enthusiasts.

All these countries also produced light-training or utility types. C.G. Taylor designed a simple aircraft, easy to fly and maintain and aptly named it Cub, and William Piper made it a success in the marketplace. The previous manufacturers of light planes had focused on the product; it took Piper's special brand of salesmanship to make the point that marketing was the key to sales. He sold airplanes not on the basis of their quality or beauty but on the bareboned fact that dealers could make money with them. Piper Cubs became the most successful light planes of the period. By this time, though, C.G. Taylor had been eased out of the business.

Piper's approach helped all his competitors, ultimately, and under the umbrella of his leadership flourished such famous names as Taylorcraft (C.G. had bounced back), Aeronca, Porterfield, Rearwin and less notable makes as the Welch, Eaglet, Funk, Interstate and others.

The period also saw the rise and fall of one beloved method of transportation and the achievement of new heights by another. The dirigible airship, favorite of so many, met the end of its commercial road, while the equally romantic flying boat came to a pinnacle of performance.

Germany had done great things with the dirigible, pri-

Evolution of Fighter Tactics: Spanish Civil War and World War II

Two schools of thought on fighter tactics emerged between the wars. Most air forces adopted rigid tactical doctrine for bomber interception, grouping their fighters by threes in tight vee formations ("vics"), or in close trail, one behind another. Tactical employment of such patterns proved effective with the help of newly introduced two-way radios in fighter aircraft and, somewhat later, by radar.

Tight formations were very vulnerable, however, if attacked by enemy fighters, since the task of close-formation flying precluded effective visual lookout to the rear. Proponents of the tight-formation school discounted this danger, though, on the theory that the range of modern bombers was too great to allow for fighter escort, so opposing fighters were not likely to meet.

The other school of thought was adopted by the German *Luftwaffe,* largely as a result of the participation of their Condor Legion in the Spanish Civil War of the late 1930s. The Germans relied on short-range bombers protected by fighter escort. Since they anticipated fighter-versus-fighter combat, they adopted flexible formations, with tactics tailored to fit their doctrine of bomber employment.

At the beginning of World War II, the Germans employed a pair *(Rotte)* of two fighters, usually flown near line abreast about 300 to 600 feet apart, allowing better visual lookout and more effective support between fighters. Two such fighter pairs formed a division *(Schwarm)* of four. They usually flew in "Finger Four" formation, the name stemming from the positions of the aircraft—similar to the fingertips of an outstretched hand. Once engaged in combat, each division would split into two sections.

This latter deployment was vindicated early in the Second World War and quickly adopted by most of the combatants.

In the most common tactic, "Fighting Wing," the leader of a two-plane section engaged the enemy while his wingman attempted to maintain an echelon position, slightly behind and to one side of the leader.

In theory, the wingman was to watch the leader's tail and warn him of surprise attacks. In practice, however, the wingman was usually so busy trying to stay in position during heavy maneuvering that he was of little use as a lookout. The wingman, indeed, often became a very easy target. Still, "Fighting Wing" provided a useful compromise by allowing an inexperienced wingman to engage in combat under the protection and tutelage of a veteran leader.

ROBERT L. SHAW

The Bracket—*(Spanish Civil War) Central Front, Spain, 1938: A pair of German Bf 109s of Franco's Condor Legion approach three Soviet I-15 biplanes of the Republican air force head-on, splitting right and left (center). The inexperienced Soviet pilots maintain their tight formation attacking one on-coming Bf 109 (upper right). The other German turns toward the formation's blind side, and opens fire.*

At dawn on August 27, 1939, the world's first successful turbojet-powered aircraft, the German Heinkel He 178, flew at the Heinkel test field in Marienhe. By the end of World War II, Germany and Britain possessed operational jet squadrons. The U.S. had test and training jets.

marily because its airship captains and crews were so highly skilled in handling the demanding vehicle. The *Graf Zeppelin* completed nine years of pioneering service, with trips across the Mediterranean and the Atlantic, the Polar regions and around the world, always in safety and luxury. The *Hindenburg,* the pride of Hitler's regime, had one excellent year in service and then fell prey to the dangers of its lifting agent, hydrogen.

Had the United States sold helium to Germany, dirigibles might have persisted for another few years. But flammable hydrogen was not the only problem. The structure of an airship is simply not strong enough to endure the twisting gusts of violent weather; in an area of thunderstorms, a dirigible more than 2.5 times as long as a football field is in mortal danger of having the front half of its structure thrust upward while equal forces push down the rear. Perhaps in the future, some epoxy-graphite structures might give the dirigible a little better chance at survival—but don't count on it.

On the other hand, the flying boat, first introduced by Glenn Curtiss in 1911, fared very well in the Thirties. During World War I, flying boats had been used extensively; after the war almost every country developed models suited to its national needs. Great Britain was by far the leading proponent of the type, building more over time than all the other nations of the world combined. The reason: its Empire needed to be connected, and flying boats had two great advantages. They offered safety over water in the event of engine failure and, more importantly, the buoyant aircraft did not require the construction of large airfield facilities. Oceans, lakes and rivers served quite well as airports, and it was easier and cheaper to build a quay than a runway.

British flying boats came into full flower during the Thirties, evolving from stately biplanes to the sleek and comfortable four-engine Empire series by the Short Brothers.

Germany, Italy and France developed distinctive types. The Dornier Wal of the Twenties continued to be improved; variants were to serve all through—and after—World War II. Italy concentrated on the twin-hull Savoia-Marchetti series, though, and like Germany produced some large and efficient multi-engine floatplanes. Japan, as it did so often, started by building flying boats under license from the British, but soon developed expertise resulting in the wartime "Mavis" and "Emily," equal to any in the world. The Japanese designations are H6K and H8K, respectively, both by Kawanishi.

But the best of all civil flying boats ultimately came from the Americans, largely because of the peculiar requirements of the American routes. While Britain's Empire boats might skip in short hops from Britain to Gibraltar to Malta to Alexandria on the way to India or Australia, the American boats had to contend with the long hop of 2,400 miles between North America and Hawaii, the longest nonstop, overwater leg in the world.

3-0-7
2-0-4

Often considered the halfway point between World War I and World War II fighters, the U.S. Army's Boeing P-26A "Peashooter" (above) was the first U.S. monoplane fighter. Left, shown with a cruiser of the United States Navy, the Curtiss SOC-1 was vital, but soon became obsolete. Launched by catapult and retrieved by crane, such floatplanes served in several navies as the eyes of the fleet before the days of radar.

The American flying boat saga started with Consolidated Commodores and Sikorsky S-40s picking their ways down the Central and South American coastlines in Pan American Airways livery. But America's "empire" was in the Far East, and the mainland had to be linked with Hawaii, the island outposts of Midway, Wake and Guam, and the Philippines.

Today a trip to Hawaii requires a routine, nonstop, adventure of five hours in a jet equipped with good radios and navigation equipment. In the mid-Thirties only the Sikorsky S-42 and the Martin 130 could accomplish this great leap. Even the S-42 could conduct only survey flights in the Pacific. But the Martin 130s—generically and universally referred to as Pan American China Clippers—could take mail and passengers as well. Only three were built—the *Hawaii Clipper, Philippine Clipper* and *China Clipper*—but they captured the imagination of the world in 1935 when one made the first flight from San Francisco to Manila.

These three pioneers (all were subsequently lost in crashes) were followed by the greatest of all the civil flying boats, the Boeing 314 Clipper (the name given in anticipation of Pan American practice.) Built upon the experience gained with both the B-17 and the XB-15 bombers (and in fact, using the

German sailors hoist an Arado floatplane aboard their ship. The Arado Ar 196A, at right, guarded the German heavy cruiser Prinz Eugen, *and now awaits its turn for restoration at the Smithsonian's Paul E. Garber Preservation, Restoration and Storage Facility near Washington, D.C.*

wing of the latter), the 314 was suitable not only for the Pacific but also for the much more demanding North Atlantic route, where weather conditions made operations extremely difficult. Forty passengers could be carried in oceanliner comfort, and the aircraft were renowned for their reliability. Only 12 were built in all, but they soldiered on through the war.

The flying boat era ended not with a bang, like the Zeppelins, nor with a whimper, but with the proud flourish of being *the* wartime transport for many of the world's great leaders. Two things caused its demise: increase in performance of four-engine land aircraft and the fact that airfields had been created all over the world during the war, making flying boats unnecessary.

The decade of the Thirties had seen an amazing effort and brilliant achievement of the industry around the world. During these years, changes in engineering and manufacturing practice and in financial methods and expectations had brought forth modern airplanes, airlines and air forces. But despite the political focus on air power, no nation had yet realized the scale of effort that would be necessary before air power became effective at a strategic level.

PILGRIM

WWII: Reaping the Whirlwind

Had the Axis leaders been chess masters and war been chess, on December 7, 1941, Hitler and company would have conceded the game. Instead, there was boasting and cheering in Berlin, Tokyo and Rome.

Real chess masters would have realized that, strategically speaking, there was no way to win. From the moment the United States entered the contest, the impoverished compact of Germany, Italy and Japan faced the combined resources of the U.S.A., the U.S.S.R., the British Empire and much of the rest of the world. Axis leaders might well have heeded Japanese Admiral Isoroku Yamamoto's rueful prediction that he could run wild for six months—but that there was little hope for Japan's ultimate victory if the war lasted significantly longer than that.

Little wonder that there was a bit of quiet cheering amongst Royal Air Force families on the day Japan bombed Pearl Harbor. It was clear that while many more British fliers would help pay for the victory with their lives, their children would not also have to die for the same cause. Whatever the cost, however long it required, the Axis powers would be put out of business for good.

In light of this all-but-absolute certainty of Allied victory, it behooves us to isolate just what role air power played in bringing about the victory. For instance, did it save lives, time, money or reduce suffering for the innocent? Why was air power—particularly strategic bombing—so ineffective for so long.

This was the supposedly all-powerful weapon that four prime ministers—Stanley Baldwin and Arthur Neville Chamberlain, Paul Reynaud and Edouard Daladier—believed would prove to be devastating to their populations in the event of war. This was the terrible arm that both Hermann Göring and Winston Churchill counted on for victory.

For most of the war, bombing proved to be an expensive means for the expression of political pique—and little more. Tactically, as support for army or navy units, air power was indispensable; strategically, as a means to smash nations into submission, it was a flop until 1944. The final, blinding stroke—the aerial deployment of the atomic bomb—still raises heated argument more than 40 years after the fact.

For Axis and Allies alike, each air arm had arrived at the war's onset, that crucial day in early September 1939, by its own route. Their relative status was affected most by national will, but also by timing.

In World War I, France had all but defined air power. Many of its planes, doctrines, organizations and leaders excelled. By World War II, that nation had lost its superiority at

Specter of dread, a Junkers Ju 87G-1 Stuka, Germany's legendary dive bomber, attacks Russian T-34/85 tanks in 1944. From September 1, 1939, the day Germany slashed into Poland, such aircraft spread terror across Europe. The Polish girl, above, mourns her sister, killed by an aircraft in a strafing attack.

every level except that of the squadron. Here, in terms of airmanship and élan, French aviators were the equal of any in the world, and Free French airmen flying for the RAF after the fall of France would prove it. At higher echelons, though, the leadership had lost the will to believe in their airmen's capabilities and permitted their fighter forces to be frittered away in support of individual army units. French bombing forces, built up at such cost, were equipped with some of the ugliest and least efficient aircraft ever assembled into an air force. Still, these could have been used at night against the Ruhr with some effectiveness. Instead, France elected not to bomb. A good show of resolve in 1939 might have called Hitler's bluff while his air strength was still low from the wear and wastage of the blitzkrieg in Poland. On the morning of May 10, 1940, the French would learn to their sorrow that Germany had not emulated their own policy of restraint. Soon it was too late to use effectively what little they had left. France would fall in just six weeks.

Between the wars, Great Britain had lurched like a somnambulist through one highly suspect government decision after another. The United Kingdom had reduced its forces to a minimum, purchased series after series of nearly identical twin-gun biplane fighters and sought to rival the French in the acquisition of goofy-looking bombers. Yet somehow, as so often, British leaders recognized the peril just in time to provide eight-gun fighters, the Spitfire and Hurricane, and to lay down specifications for such effective bombers as the Wellington, Halifax and Lancaster.

It was as if those great heroes of the Napoleonic wars, the Duke of Wellington and Admiral Horatio Nelson, had been looking things over from above (giving them both considerable benefit of the doubt), and had reached down to give crucial guidance from 1934 on. By September 1939, Britain had provided itself with a small air force equipped with some highly promising types. Within the year, a bustling industrial buildup would exceed German capacity in every quarter—airframes, engines and equipment.

In 1934, Stanley Baldwin had perceived at last that the

France, 1940: Royal Air Force (RAF) pilots of No. 87 Squadron, left, sprint toward their Hawker Hurricane fighters. Several aircraft have been equipped with three-bladed propellers for higher performance. After British units were forced to abandon France in June 1940, "Chain Home" radar stations, below, gave vital early warning of German air raids.

"frontier of Britain was not at the Channel, but on the Rhine." Then, Britain's air arm consisted of mediocre wood-and-fabric biplanes. Five years later, and not a moment too soon, the British Isles possessed a force of modern all metal aircraft and a subsystem of shadow factories blossoming almost unseen within ancient manufactories across the land. Perhaps most importantly, for the first phase of the battle the planners set Sir Henry Wimperis, Sir Henry Tizard and Sir Robert Watson-Watt to create almost from whole cloth the implausible, mysterious shield called radar.

Even the arcane British political system had worked to the advantage of the RAF. Churchill, one of the men most responsible for the infamous "Ten Year Rule," which advocated moderation in armaments, had by the late Thirties become one of the most vociferous proponents of rearmament. And Neville Chamberlin, that Chester Gump of politics, managed through no fault of his own to buy a year for the British air force at Munich in 1938. (The question of whether this was to Britain's—or the world's—overall benefit is often argued; there can be no question about its substantial benefit to Britain's air force.)

But it was in leadership that Great Britain really shined. It was leadership at all levels—from Air Chief Marshal Sir Cyril Newall, Chief of the Air Staff until 1940, down through the ranks to the corporals and airmen. It was a splendid chain of command, a superb fabric of motivation and discipline.

There had not been much in the way of funding during the interwar years, but Air Marshall Sir Hugh Trenchard and his successors saw that what money there was went for training, for the establishment of bases and for the maintenance of adequate supply methods. He and his policies kept the organism of a true air force alive, and when expansion came, Britain possessed a hard core of experienced leaders at all levels to make the best use of its fledging air force. And for good or ill, it was the leadership that persisted in the allocation of resources to Bomber Command in the face of its grievous losses and its apparent lack of success in strategic bombing up until mid-1944.

Germany's image, carefully fostered by those conducting visitors through German factories, was one of total precision and order. Reports circulated of complete aircraft factories ready to enter production, so totally equipped that on each empty desk lay three precisely sharpened pencils—one hard, one medium, one soft.

In effective organization, Germany's strength lay somewhere between that of Britain and France and just behind Britain in the creation of modern aircraft types. German engines were generally superior to those of the French and inferior to those of the British until, of course, the deployment of jet fighters. Surprisingly, and contrary to the stories floating about before the war, the general quality of German aircraft was very high, especially in areas where it counted. A proud tradition of manufacturing excellence virtually guaranteed this result. To the very end, German craftsmen and mechanics gave their best effort for their nation's lost cause.

Despite the cultivated image of ordered reason, of "rationalized industry" and huge, overdesigned factories deliberately laid out to minimize bomb damage, Germany was ill-prepared for a major air war. Manufacturers hoarded supplies of scarce materials; most factories worked on a single shift and closed on weekends. Germany was not serious about the war, had not truly mobilized, had not suspended

PR F
PR W

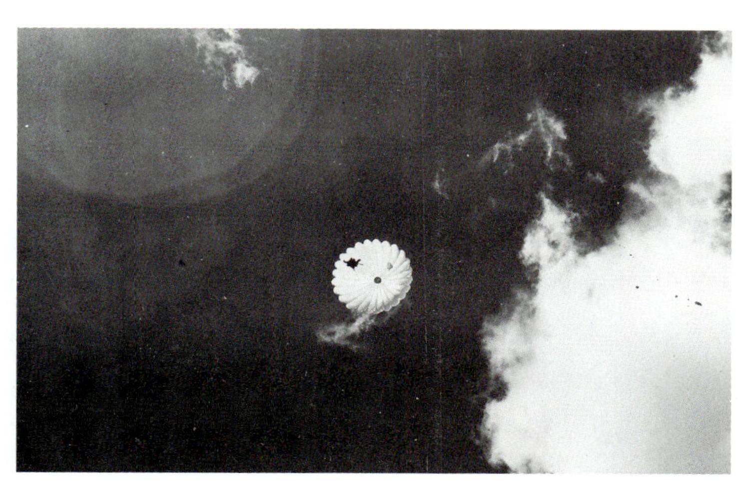

production of purely peacetime civilian commodities and had made little effort to bring women into the workforce to replace conscripted men. In fact, women were encouraged to stay in domestic service, to patronize beauty parlors, to do everything possible to maintain a peacetime facade to foster the illusion of butter with the guns.

In terms of quantities of aircraft, bombs, searchlights, fuel tankers, flak units—everything—Germany set its sights far too low. When production reached about 700 aircraft a month in 1939, planners deemed this level adequate for the war they assumed was going to be a short one. By 1944, when it was too late, the rate would reach 4,000 per month. They made essentially wrong decisions such as producing twin-engine medium bombers rather than four-engine heavy bombers. This fit perfectly with their doctrine of tactical employment. Here as elsewhere, the decision makers pulled their punches, often changing their minds about priorities. For instance, the Blitz of London found bombers built for tactical missions used in a strategic role, for which they proved inadequate.

But the real difference was in leadership. In this, Germany clearly lacked Britain's greatness. The Treaty of Versailles, which ended World War I, had destroyed the framework of the German Air Force—not all bad, as it was an awkward organization. Nothing had developed in its place prior to the clandestine reestablishment of the Luftwaffe after 1934.

General Hans von Seeckt, chief of the German army in the Twenties, had done as well as possible given the limitations of the 100,000-man Wermacht. In fact, many of his choices for the Luftwaffe leadership proved first-rate. We have but to

Royal Air Force Supermarine Spitfires attack German Messerschmitt Bf 109 fighters in this Frank Wootton painting of action on the morning of September 15, 1940, the climactic day of the three-month-long Battle of Britain. Victorious British pilots retained control of the air and forestalled a German invasion. Above, a parachuting airman descends toward a rescue vessel.

Overleaf: Achtung! Spitfeuer! *warned German pilots as Britain's premier World War II fighters hurtled at them. Though Britain possessed greater numbers of sturdy but slower Hawker Hurricanes, the nimble Spitfire proved to be a match for the versatile German Messerschmitt Bf 109. Carefully restored, these "Spits" display the clean lines designed by Reginald Mitchell in 1935.*

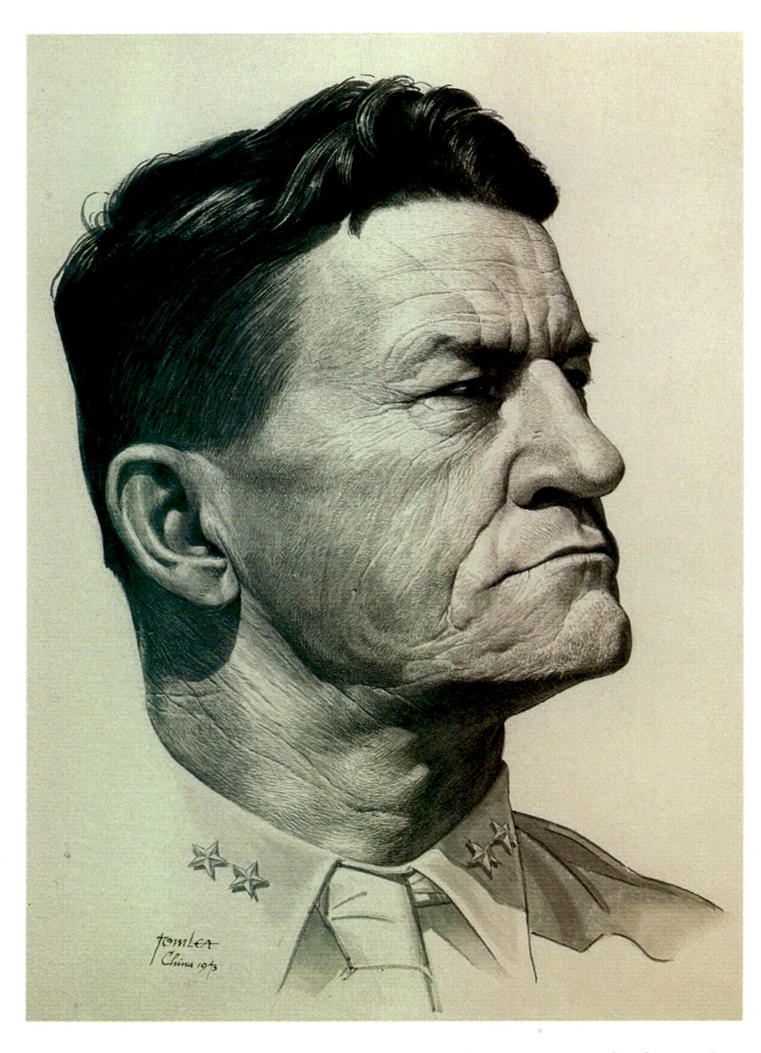

Artist Tom Lea caught the steely determination of Maj. Gen. Claire Chennault in this 1943 portrait. Months before Pearl Harbor, Chennault formed the American Volunteer Group, the "Flying Tigers," to battle the Japanese in China. Opposite, December 12, 1943: a Japanese Nakajima Ki-43 "Oscar" fighter goes down after colliding with a USAAF Curtiss P-40N piloted by Flying Tiger Fighter Group Lt. Donald S. Lopez near Hengyang, China. Though missing three feet of its wing, Lopez's sturdy Warhawk, Lope's Hope, *brought him safely back to base.*

remember General Walter Wever and Hans-Jürgen Stumpff, Albert Kesselring—later a field marshal along with Erhard Milch—from Lufthansa, the national airline. Yet the Reichsluftwaffe expanded so fast after 1935 that there was little chance to revitalize the chain of command, to train capable commanders of the highest quality. Despite the image of cold Nazi efficiency, in terms of supply management, radiotelephony and similar elements there was a tremendous lack of discipline within the Luftwaffe. Nothing like this would have been tolerated in Britain.

Luftwaffe chief Hermann Göring later sought to work organizational miracles through the rapid promotion of successful young fighter pilots to command positions. He succeeded with such men as Adolf Galland and Werner Mölders. Control of air operations is one thing; staff work—seeing to replacements, spares, training, discipline, fuel supplies and movements—quite another. Not all great fighter pilots make great commanders. The Germans tried to buttress the system with experienced staff officers—usually World War I veterans—who handled as much of the paperwork as possible. But the commander is always the commander, no matter who is backing him up.

After the Germans invaded Russia in 1941, the fighting fronts expanded and attrition increased. The Luftwaffe soon ran head on into some hard facts of life and death. Naval fleets sortie periodically, and divisions on the ground engage and then disengage the enemy, but the airplane is flung into battle every day that it is airworthy. It came as an unwarranted surprise that the air battle was continuous—the Luftwaffe allowed its units no respite, grinding men through combat like berries expressed through a colander.

Little wonder that the Luftwaffe's command structure became unglued at the peak of its expansion, when Germany was putting pennypacket units into Africa, Norway, all along the 3,000 miles of Russian front, in Italy—all as the Luftwaffe gained responsibility for the air defense of continental Europe. When the war situation changed, when Germany was compressed again within its borders, the dying Luftwaffe was squeezed into a compact mass. Not planning but the adverse fortunes of war brought together again enough leaders, supplies and airplanes (if not fuel) for operational efficiency.

The Italian opéra-bouffe situation was as might have been expected in Benito Mussolini's lukewarm Fascist regime. Some fine factories produced aircraft suitable for the civil war in Spain or the invasion of Ethiopia, but not for modern warfare. *Il Duce*'s realm was short on supplies, enthusiasm and expectation. Even at the highest levels, dissatisfaction and disloyalty prevented effective rearmament. The Fascist leadership was rife with privilege and injustice. Though morale ran high in the flying squadrons and the pilots were excellent, they took little delight in fighting the war as the Luftwaffe's lackeys. Where else could a man of such international reknown as Air Marshal Italo Balbo be shot down by his own antiaircraft fire—flak that rarely damaged the enemy? Where else could a craft like the Macchi C.205 be produced in such tiny numbers? A Ferrari of a fighter, perhaps the match of the Spitfire, it was produced in numbers fewer than 400, compared to the 22,000 Spitfires of all marks, the 14,000 Hurricanes and 15,489 Mustangs, just to begin the roll call of Allied aircraft production. Allied pilots in occupied Italy vied for the pleasure of taking a turn in a captured Macchi. If beauty and mechanical excellence were the deciding factors in warfare, Italy might well have conquered the world.

The United States was to be blessed with a two-year delay before entering the war, two years in which the infusion of orders from Britain and the knowledge gained from observation would make all the difference in the world. The greatest difference between the United States and other countries, of course, was in American industrial capacity. The U.S. aviation industry, so laughingly discounted by both friends and enemies, was producing military aircraft at a rate of about 2,200 a year in 1939. By 1944 the United States turned out planes at a rate of nearly 100,000 a year, some of them the era's most complex machines.

The same two years of grace permitted an expansion that preserved and supported the strong leadership that had emerged between the wars. Much of this was due to the American emphasis on cooperation and teamwork—factors that contributed to effectiveness of leadership at all levels. Americans were able to pull together in a way that overcame rivalries among the services. In a sense it was easy; the men who had served as first lieutenants or majors, ensigns or lieutenants, for years saw the expansion as having sufficient promotion potential for everyone. President Franklin Delano Roosevelt's promise of 50,000 aircraft a year offered enough for all, especially to services that had purchased aircraft in units of a dozen or less at a time.

Left, Japanese A6M2 Model 21 Zero fighters warm up for take-off. The scene may be the flight deck of carrier Zuikaku *just before the December 7, 1941, attack on the U.S. Pacific Fleet at Pearl Harbor. Above, in this painting of the attack by Robert McCall, a Zero pilot, his canopy open, flashes past Aichi D3A1 "Val" dive bombers wheeling over "Battleship Row."*

". . . a date which will live in infamy . . ." declared President Franklin Delano Roosevelt of the devastating Sunday morning attack that took the lives of more than 2,000 U.S. servicemen and civilians. Americans remembered forever the moment they learned of the surprise raid.

The Axis leaders also underestimated the abilities of the Soviet Union. Hurt by purges, ill-trained, saddled with stolidly doctrinaire tactics, this tortured giant possessed more resources than anyone could imagine. Though the Soviets began the war with few modern aircraft, they had undertaken an industrial expansion plan similar to that in the United States. While the equipment varied in quality, the overall standard matched that of most European nations. Russia's greatest strength, of course, was in its traditional capacity for absorbing punishment and its equally resilient potential for recovery. The Axis powers simply had no idea of the geographical scale, industrial strength and popular determination with which they had tangled.

Japan should have known better, too, especially since it was locked in combat with China. The makers of Japan's military doctrine created small, highly skilled army and naval air forces solely for the support of their respective services and which could work together only with difficulty. Yet there were perhaps no finer combat pilots in the world than the Japanese—the product of a national obsession with quality that expressed itself in training and doctrine.

But they bought their combat edge on the cheap, against unprepared opponents. Victory after victory led the Japanese to believe that they need afford only a small air force, well-equipped, highly trained and superbly led. They knew that they would prevail. Prevail they did—until they ran into a *huge* air force, well-equipped, highly trained and superbly led. When that happened, the small and irreplaceable air forces of Japan were consumed by fire. Admiral Yamamoto's strategic assessment of a war forced on the Americans was correct; nowhere was his counsel more pertinent than for the air war. On the day of the Japanese attack against Pearl Harbor the Japanese were infected with "the victory disease," and it lead to the demise of the air arm.

Hubris among the Axis leaders as early as the autumn of 1939 was surely understandable. After a string of diplomatic triumphs that won vital concessions and vast amounts of territory, the Germans pulled off a dazzling political coup with their Russian peace pact of late August 1939. The Junkers Ju 87 Stukas, Messerschmitt Bf 109s and Heinkel He 111s established air superiority over Poland, and worked in perfect concert with the rapidly advancing army units. If not bedazzled by the military performance, Germany's opponents on the Western Front were mesmerized by the most appallingly fatuous propaganda in the history of warfare. The Allies went on to manage the most supine acceptance of entrapment imaginable.

But what else could they do? After spending hundreds of millions of francs on their air forces since 1918, the French found themselves with *no* air power. Of about 4,300 aircraft on strength, only about 1,000 were modern, and of these only 700 were complete. The British had some 5,711 "first-line" aircraft, of which 1,911 were of genuinely first-line quality, a mixed bag whose character changed for the better month by month with the infusion of newer types. The Germans had 7,402 aircraft, 3,609 first-line.

But this told only part of the story. British output was beginning to rise at a remarkable rate; from 2,827 planes in 1938 to 7,940 in 1939. By 1940, British manufacture of aircraft had surpassed Germany's and would remain ahead until Minister for War Production Albert Speer's fitful spasm of war-industry activity during 1944. In France, aircraft production was clearly inadequate to place that nation on a war footing and, in any case, ended in June 1941.

Though prompt, the first Allied combat sorties were clearly disappointing; the Blenheims sent against the German fleet at Wilhelmshaven within 24 hours of the war's beginning hit few of the anchored ships, and the bombs that did failed to explode because of improper fuse settings. The raid characterized the beginning stage of the war: overoptimism, too much consideration of the German civil populace, woefully poor targeting, inaccurate drops and, of course, dud bombs. A decision was made not to attack the Ruhr, the heartland of German armament, in part because "private property might be damaged."

The Allies watched Germany and Russia divide Poland, and then felt all of their apprehensions come true as Russia gobbled up the Baltic states and began to bully Finland. Incredibly, with the exception of the limited "Saar" offensive, the Allies thought of bombing Russia, invading Norway, occupying the Middle East, anything—everything but confronting the Nazi Army in open battle on the Western Front. During late 1939 and early 1940, the period of so-called Phony War, the Allies waited.

Hitler did not. In a series of brilliantly executed, highly risky strikes his armed forces conquered Norway, Denmark, Belgium, Holland and France. The Blitzkrieg was rolling again, but the first sign that the Luftwaffe was not in fact invincible occurred over Dunkirk as the British extracted their deflated army from France. Everything prior to that had been according to form, from the Luftwaffe strike of May 11 against 70 French and English air bases, to the clockwork precision with which Stuka dive bombers were summoned to crack defensive positions. Obsolete British bombers were consumed in futile attacks against German armor, shot down either by mobile flak batteries or the roving Messerschmitts. The French dug deep into their bag of ancient aircraft and dispatched Loire floatplanes and Amiot night bombers. Flying better planes and experienced from action in Spain and Poland, the veteran German pilots enjoyed a turkey shoot.

With the Battle of France all but over and Britain at its most vulnerable, British Air Chief Marshal Sir Hugh Dowding refused French demands for continuing air support; he held back his precious fighter squadrons from the continental conflagration, and Churchill backed him. This made all the difference a few weeks later in the Battle of Britain.

The autumn of 1940 saw all of the weaknesses of the German system come up against all of the strengths of the British—and still the Battle of Britain was a close-run thing. The British had bought their air force with the closest possible eye to economy; had they spent only a few hundred thousand pounds less, or spent them later, Germany would have won the Battle of Britain and invaded.

Just a few pounds less on base expansion, just a few pounds less on Spitfires and Hurricanes, just a few pounds less on shadow factories and, most importantly, just a few pounds less on the development of radar and the early-warning system, and Britain would have been overwhelmed. It was a case where the politicians had inadvertently obtained just enough pow for the pound, where the barking, insistent demands of Churchill and his followers had been just enough, just *exactly enough* to carry Britain to the point where she could survive.

And the Germans suffered in the opposite direction. Had the Germans spent just a little more to develop four-engine bombers, had they spent a little more on their own radar development (at least so that they would have recognized what the British were doing), had they taken the care to equip

On April 18, 1942, the Doolittle raid against Tokyo and other industrial centers stunned the Japanese and cheered the Allies. Led by Lt. Col. James H. Doolittle, already famed for his pre-war research and racing flights, the raid consisted of 16 Army B-25 Mitchell medium bombers launched from the aircraft carrier U.S.S. Hornet. *All aircraft reached their targets; nine like Capt. C. R. Greening's airplane, pictured at left by artist Keith Ferris, were attacked by enemy aircraft. Greening and his crew made it to China where they bailed out.*

their fighters with long-range, droppable tanks (an unforgivable oversight, considering this device had been around for a decade), had they not so politicized their intelligence system that it gave unusable information, had they possessed a Dowding running the Luftwaffe rather than a Göring, then they undoubtedly would have won.

Hitler's odd combination of the parvenu's respect for the British Empire, distrust of the sea and faith in the ancient Teutonic lust to smash Russia—Communist or otherwise—worked in Britain's favor. The Blitz itself revealed that the Luftwaffe's means were inadequate to defeat England from the air. Neither the airplanes—mostly Heinkel He 111s, supplemented by Dornier 17s and Junkers Ju 88s—nor their navigation and bomb-aiming systems were adequate. The Ju 87 was quickly withdrawn from operations, and the twin-engine Messerschmitt Bf 110 fighters sorely needed escort by the short-range Bf 109s.

The Luftwaffe could not win the war in the air, the RAF could not be eradicated. Above all, Britain would not be invaded. The Blitz, the daylight bombing of London, dwindled during the rest of 1940; people were still killed, and no one knew when the danger would be renewed, but the first and crucial Allied victory was won. Throughout the rest of 1940, the air war gradually spread to other fronts. It was typified by Germany's inappropriate tactics, obsolete equipment and, as always, a basic shortage of will and timing.

Italy's entrance into the war in June 1940 opened the possibility for conflict throughout the Middle East, where "colonies" of England and Italy butted against each other. British Air Commodore Raymond Collishaw (an ace with 60 victories from the first war, former commander of the famous "Black Flight" of Sopwith Triplanes and veteran of fighting in

Warned by radio code breakers of Japanese fleet movements, U.S. forces won a pivotal victory at the Battle of Midway, June 3–6, 1942, destroying four enemy aircraft carriers. Above, artist R. G. Smith pictured U.S. Navy Douglas SBD Dauntless dive bombers passing mortally stricken Akagi. *Left, a naval aviator peers into an aiming device as he dives toward a target.*

As men went to war, women did also, many working on aircraft assembly lines like this Consolidated B-24 Liberator plant at Fort Worth, Texas. U.S. industry produced an astounding total of more than 360,000 aircraft during World War II.

Russia in 1919) sent his squadrons into action immediately against El Adem in Cyrenaica (northern Libya).

The British could field 29 squadrons totaling 300 aircraft for Egypt, the Sudan, Palestine and Transjordan, East Africa, Aden and Somaliland, Iraq and adjacent territories, Cyprus, Turkey, Yugoslavia, Rumania, Bulgaria, Greece, the Mediterranean Sea, the Red Sea and the Persian Gulf, an area of 4.5 million square miles. What a glorious hodgepodge of aircraft they were! Eighteen squadrons had reasonably modern equipment—Blenheim I bombers, Sunderland flying boats, Lysander observation planes and Gladiator fighters. The rest included such various, obsolete types as Bristol Bombays, Vicker Valentias, Wellesleys, Vincents, Battles, Junkers Ju86s of South Africa's air force, Hardys, Audaxes, Harts, Hartebeestes and Londons—all antiques with wings.

The Italian opponents were perhaps better equipped, with Fiat's C.R.42 biplane fighter, equivalent of the Gladiator, and the Savoia-Marchetti S.M.79 Sparviero bomber-transport. Also, the Italians could reinforce quickly from the mainland. The widely spaced areas of conflict were soon complicated by the surprise attack on Greece by Italy on October 28.

Because of the aggressiveness of the British RAF and the tenacity of the Greek Army, 1940 ended upon an implausibly cheery note. Victories in Cyrenaica enabled the British to help Greece defeat the Italians. Both successes spelled disaster for the British during the next year, for they invited the replacement of Italian arms with German ones in both Greece and North Africa.

In Japan, the Nakajima Ki-43, the "Oscar," the Mitsubishi A6M1 Zero, or "Zeke," and the G4M "Betty" all made their first flights during 1939 and 1940—both vintage years for today's aircraft buff. The Allies learned little of Japanese advances until much later, of course, but were well aware of the debut of America's Lockheed P-38 Lightning, Consolidated B-24 Liberator and the Boeing 314 Clipper. These fledglings would mature in combat two years later when America entered the war.

Germany witnessed the prototype Focke-Wulf Fw 190's first flight, uncertain if the fighter could be brought into production before the war would be concluded by victorious Axis forces. The flights of the Heinkel He 176 rocket plane and the He 178 jet, while experimental, were quite interest-

Navy Secretary Frank Knox inspects North American SNJ advanced trainers awaiting their naval aviation cadets at Corry Field, Pensacola, Florida, in 1941. Such schools trained hundreds of thousands of Allied aviators and crews.

ing. A huge bomber, the Heinkel He 177, started an ill-fated career. In England, the Halifax, Beaufighter and Stirling also flew for the first time.

If 1939 was good, 1940 was even better. Still neutral, Russia introduced the Yak-1, MiG-1 and LaGG-3 fighter prototypes and, according to Stalin, more important than any other, the Ilyushin Il-2 Stormovik ground-attack type. More than 36,000 of the latter would be built to crush the Nazi armored shield. Britain flew the Hawker Typhoon and de Havilland Mosquito prototypes; from the United States came two fighters of renown, the Vought F4U Corsair and the North American P-51 Mustang, the aerial razor blade that would slit Göring's throat over Berlin, and the Martin B-26 Marauder bomber. Italy electrified the world by flying a propellerless airplane in public, the hybrid Caproni-Campini, a jet with a compressor powered by a piston engine. It hurtled along at 130 miles an hour.

Though revolutionary, Germany's development of the turbojet engine didn't pay off. Again Germany's research and development efforts were half a war too late. The Henschel Hs 293 radio-controlled glide bomb flew, and the DFS 194 research craft, which grew into the Messerschmitt Me 163 fighter powered by a rocket motor. Hitler's new weapons were amazing indeed, but not powerful enough or sufficiently plentiful to turn the tide of war in his favor.

Only in retrospect is it possible to call the 12 months between mid-1941 and mid-1942 the turning point for the Allies, for during this crucial year they sustained defeat after defeat. Efficient, well-led and daring German forces, acting in combination, overwhelmed the Allies on every front. The Germans pushed the British out of Greece and Crete in evacuations that lacked even Dunkirk's dim claim to a bright side. The Balkans fell to armored columns faster than a private car could traverse the same area. Russia seemed to reel toward defeat. In the Orient, American General William "Billy" Mitchell's predictions of United States vulnerability had come true with a vengeance at Corregidor, Bataan and as far west as Burma. Japan seemed unstoppable.

Yet the high-rolling Axis gamblers had lost; they were dead and didn't know it yet. For they had forfeited any chance of a negotiated settlement and did so without any clear idea of what might follow. Diversions in Greece and the Balkans had

somewhat delayed the German timetable in Russia; but how could a country that had sharpened pencils in its unused aircraft factories (one hard, one medium, one soft) send armies into the heart of Russia without winter clothing, antifreeze or alternatives? How could such a country, tightly run from the top by "one genius," have so much difficulty in increasing production that its maximum strength could decline from 3,692 aircraft in March 1940, when it faced only France and the British Commonwealth, to 3,451 in June 1941? By then, Germany was engaged with Russia as well and had spread itself thin over much of Europe and North Africa. The answer of course lies in the fact that a genius was not running the country but rather the same sort of military bureaucracy that had malfunctioned for General Erich von Ludendorff in World War I. Whatever early genius Hitler displayed in bluff and blitz, his momentum failed in the East as in the West.

So successful over Poland and France, the Luftwaffe found itself on the ground in Russia—and in grievous trouble. Cold weather either encased airplane tires in frozen mud or cracked the rubber made brittle from freezing, kept engines from starting, stalled support vehicles, forced men to huddle together in trenches to delay but not avoid freezing to death. Aircraft mechanics were more fortunate. They had machines designed to warm engines for flight, and they could employ the heat to unfreeze their hands from metal tools and parts. Such unauthorized use of equipment allowed aircraft engines to freeze up.

Once again it was a question of scale. Hitler's Luftwaffe was thrown into Russia just as a handful of sand is thrown into the wind. And once again, initial success was extracted from the blood of the few—often fanatic—pilots who managed to fling themselves into the air again and again—not from an integrated chain of command. The men on the line were never backed up by reliable supply channels. Even as ice ensnared the Luftwaffe, the Russian Air Force began its recovery from the disastrous summer of 1941.

The Japanese had even less reason to anticipate their doom. Exalted from inhaling the smoke of Pearl Harbor and the Philippines, from having sunk HMS *Repulse* and HMS

The Lockheed P-38 Lightning, opposite, was the only American fighter in continuous production from before the war until the end of hostilities. The long-range P-38 served in every theater in a variety of capacities and was credited with shooting down more Japanese aircraft than any other type.

On April 18, 1943, 16 USAAF P-38 fighters flew more than 550 miles from their Guadalcanal base to the island of Bougainville to intercept a Mitsubishi G4M "Betty" bomber carrying Admiral Isoroku Yamamoto, below right, master strategist of the Japanese navy. Yamamoto died in the jungle crash, below left.

Prince of Wales as a prelude to gobbling up the rest of Southeast Asia, they looked forward to victory in Australia, India and Ceylon. All their operations had gone off with unparalleled success; they were within a few months of establishing a defensive perimeter that would secure supplies of oil and rubber. As Japan saw it, the United States had no other choice but to negotiate. No one in Japan asked what might happen if America declined this honor.

Bravery and dash distinguished air combat throughout the year, nowhere more so than during the RAF's valiant defense of the Mediterranean island of Malta. The Germans and Italians might possibly have taken this "unsinkable aircraft carrier" by surprise, thus sealing the fate of the Mediterranean and probably North Africa. Once again the British hung on, dug in and ultimately prevailed.

The RAF began pecking at targets in Europe, just as Germany continued its greatly reduced night bombing raids against British cities. The British launched Operation Rhubarb, their daylight fighter sweeps over France. In May, the Germans carried off a superb airborne attack that secured Crete, though it was a Pyrrhic victory. Attention was diverted momentarily from the Cretan debacle when a Catalina flying boat spotted the German battleship *Bismarck* in the North Atlantic. Torpedo-equipped Swordfish from the carrier H.M.S. *Ark Royal* crippled *Bismarck,* the greatest threat to the Royal Navy in the German fleet, permitting surface forces to finish her off.

In September, German pilot Hans-Ulrich Rudel, the most decorated man in the Luftwaffe, sank the Soviet battleship *Marat* with a single 2,000-pound bomb from his Ju 87 Stuka.

The British Fleet Air Arm carried off perhaps its most brilliant strike at the Italian harbor of Taranto, surpassed in naval warfare only by the success of the Japanese attack on Pearl Harbor. Ancient Swordfish, the beloved "Stringbags," went in and smashed three battleships, transforming Mussolini's "Mare Nostrum" (the Mediterranean) into a mere "Nightmarum." Late in the year, the RAF's American Eagle squadron, composed entirely of volunteer American pilots, went into action over Northern France, evoking memories of the Lafayette Escadrille of World War I.

The Jug in New Guinea

Edwards Park

I work standing up. When people ask why, I murmur something about my back. The real reason goes back to December, 1943, when the Pentagon rediscovered our long-forgotten fighter squadron in New Guinea's Ramu Valley, and gave us P-47s.

We had been flying P-39s, dicey by nature, and growing more so with advancing age. So my first memories of the Thunderbolt are tinged by obvious contrasts: It seemed huge to pilots used to the tiny '39. "Hey, we got Cadillacs," one colleague remarked—and searched the spacious cockpit for a glove compartment.

Seated in elbow-spreading comfort, we checked out. Takeoff? You checked mags, set trim tabs, cracked flaps, locked tailwheel, poured on coal and the plane did the rest. It obeyed every whim—climbing steeply, abruptly winging over and diving—without a sign of P-39 hysteria. On landing, it simply sagged impassively, straight onto the runway. Unused to 16,000 pounds of airplane, we sometimes clumped it on a little hard. "I just dropped eight tons on the Nadzab strip," a fellow pilot once ruefully reported, "and I was in it!"

No matter. The P-47 was built like a boulder. I once saw one touch water on a low pass, and fly home with a foot of each propeller blade curled neatly back. Another collided with a P-38. Both got back, the Lightning with half a wing, the Jug with scratched paint. It flew another mission that afternoon.

Those missions were far longer than any we had flown before: four, even five hours. But even these missions weren't long enough for the Southwest Pacific. So Charles Lindbergh, representing Pratt & Whitney, briefed Fifth Air Force Thunderbolt pilots on long-range flying. I remember gathering in a lantern-lit mess tent one evening, and being introduced en masse to the god-figure of all pilots everywhere, tall, slender, familiar. Someone whispered that he had shot down an enemy plane that afternoon. Of course he never mentioned it. In his soft Midwestern voice, he told us to try a high throttle setting and only 1,400 revolutions per minute. Our propellers would grind along, taking great bites of air.

We stirred uneasily at the heresy, and Lindbergh read our thoughts. "These are military engines, built to take punishment," he said. "So punish them." Then he added that if we felt uncomfortable about flying this way, we shouldn't do it: "You're the captains of your own ships. You must make the decisions. After all, you know more about flying your planes than I do."

At that, we burst out laughing. And we began flying eight-hour missions. We'd return so numb that we'd taxi standing up, feet on rudder bars, massaging our buttocks with both hands.

Oh, yes. The P-47 was wonderfully forgiving and brutally tough. And thanks to it, and to Charles Lindbergh, I now work standing up.

All aluminum fuselage gleams in the Keith Ferris painting of a Republic P-47D Thunderbolt—an aircraft similar to the one flown by Edwards Park over New Guinea. This largest of all WWII single-engine fighters (weighing 8 tons) possessed a 13-foot propeller, a 2,300 horsepower Pratt & Whitney engine and a top speed of 428 miles per hour.

The United States increased production of new types, incorporating lessons learned in the first two years of the war. It built such models as the Republic P-47 Thunderbolt and the Grumman TBF Avenger (which future Vice President George Bush would fly in combat). Before the end of 1942, the United States had flown the prototypes of all the planes that it would produce in such astonishing numbers through the end of the war. These included the legendary Chance Vought F4U Corsair and the North American P-51 Mustang. Britain flew the prototype Avro Lancaster night bomber that would help grind Germany into the dust, as well as the Gloster E.28/39 that initiated Britain's jet age.

Japan fielded the four-engine H8K "Emily"—considered by many the war's finest flying boat—as well as the Kawasaki Ki-61 "Tony" and the Nakajima J1N1 "Irving." (Such names were assigned by the Allies; the Japanese, of course, had their own nomenclature.) Germany put several oddities into test, including the Heinkel He 280V1 jet fighter, the Messerschmitt Me 163 Komet rocket-powered interceptor and the five-engine, twin-fuselage glider tug, the Heinkel He 111Z. Russia put out the Lavochkin La-5 fighter.

As terrible as 1941 had been, the early days of 1942—the war's third year—were far worse. Japan inflicted the most humiliating defeat in history on the English at Singapore, then proceeded to acquire Java and Rangoon. Hitler, apparently completely recovered from the rebuffs of the Russian winter campaign of 1941, moved southeastward, perhaps intending to link up with the Japanese in India. Lt. General Erwin Johannes Eugen Rommel began his dazzling exploits in the North African desert, swallowing Tobruk at a gulp and then setting his forces swarming towards Alexandria, whose fall might have been the Middle East's version of Singapore. Here were classic Axis advances and vintage misfortunes for the Allies.

Yet everywhere the Axis leaders made strategic miscalculations of the highest order. All the while the Allies stolidly hunkered down to the business they knew best: mass con-

Top-scoring American aviator of World War II, Army Air Force Lockheed P-38 Lightning pilot Richard Bong, above, shown here in New Guinea, shot down 40 Japanese aircraft. Top, U.S.S. Lexington *pilot Alexander Vraicu shot down five Japanese airplanes in eight minutes in his Grumman F6F Hellcat during the "Marianas Turkey Shoot," June, 1944. The Japanese lost 366 aircraft while only 26 Americans went down.*

Baracca

The best Italian fighter of the war, used mainly in North Africa, the Macchi C.202 Folgore possessed excellent handling qualities and a 375 mile-per-hour top speed. Out of more than 1,100 produced, this aircraft, displayed in the Smithsonian's National Air and Space Museum, is one of only two left.

Inset, headquartered in a prewar North African casino, U.S. Twelfth Air Force intelligence officers plan a January 1943 raid. LIFE magazine's Margaret Bourke-White photographed the session before climbing into a B-17, becoming the first woman to accompany a World War II American bombing mission.

scription, mass production and the taking of the war into the foe's backyard. An early glimmer of the last factor was Lt. Colonel (later General) James "Jimmy" Doolittle's bombing of Tokyo—not so much a bombing raid as an engagement ring—a promise of things to come.

Then the future was writ large in the Coral Sea, where the Japanese and Allies fought, for the first time in history, a major naval engagement in which the surface ships never sighted each other. A month later near the U.S. territory of Midway Island—west of Hawaii—an American carrier task force commanded by Rear Admiral Raymond Spruance won a victory during history's second standoff carrier battle. The lesson of the first in the Coral Sea had been learned well. Furthermore, the Japanese never realized just how much that superior American electronic eavesdropping contributed to wrapping a shroud of defeat around Japanese hubris. Douglas SBD Dauntless dive bombers destroyed four Japanese aircraft carriers at the Battle of Midway. The Japanese were never to regain the strategic offensive; from that moment on, the Imperial war machine was forced to shore up dead strategy with dead soldiers, sailors and airmen.

In Africa, an uninterrupted series of Allied victories began at the Battle of El Alamein near Alexandria, Egypt, where tactical air power and superior armored strength and artillery gave General Bernard Montgomery overwhelming superiority over Rommel's forces. These were the kind of house odds that he always insisted upon. In Russia, history was about to repeat itself for Hitler. Totally underestimating the power of the Soviets, he entered into a series of strategic errors which led, inexorably, to stunning defeat at the horrendous Battle of Stalingrad. This Soviet victory, with that of America at Midway, are widely recognized as the war's two great turning points.

Words like élan, willpower, leadership, moral superiority and other intangibles hone the blade of war but do not constitute its heft. In the last analysis, the air force with the most and the best planes will always win. This fact dominated Allied plans.

During 1942 the United States alone produced 47,836 aircraft, twice or more that of the Axis. The proportions were almost exactly the same as the percentage figure of world manufacturing capability in all categories, in which the Allies controlled 60 percent, the Axis 17 percent and the neutrals, most of whom favored the Allies, the rest.

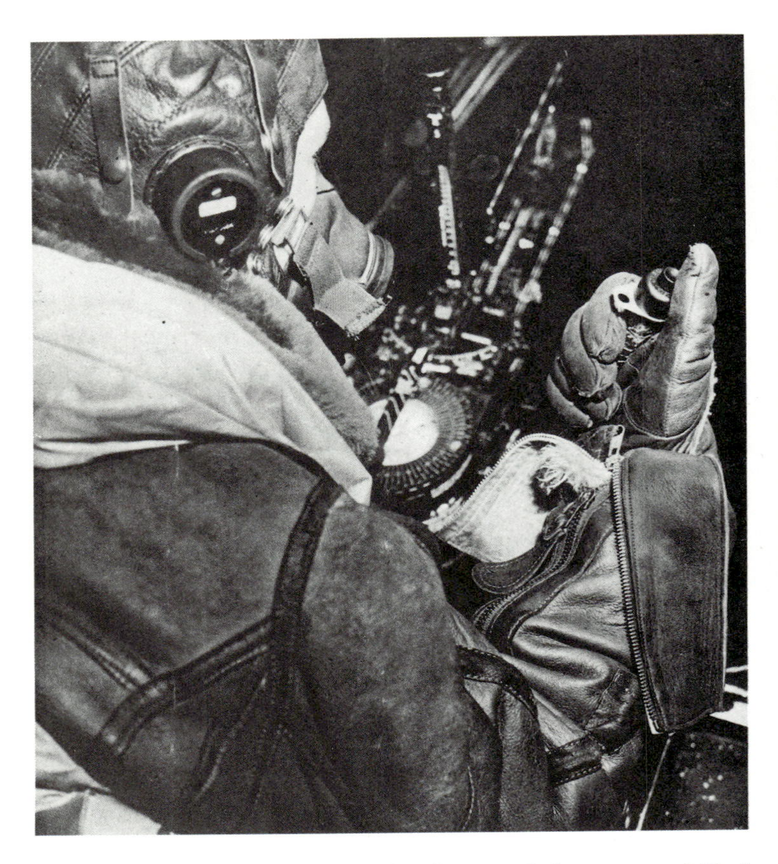

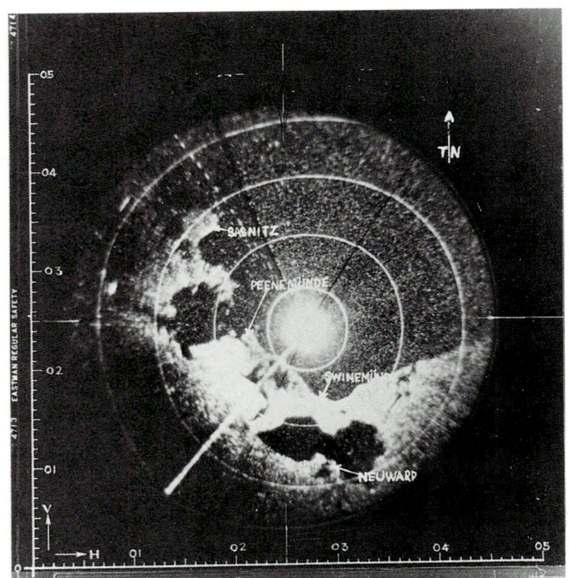

While the USAAF bombed by day, Britain's RAF harried the Axis by night. Guided by H2S airborne radar, above, called "Mickey" by Americans, British bombardiers, left, hit such targets as Peenemünde, the great German rocket development center on the North Sea, seriously delaying Hitler's V-2 missile program.

The Axis powers suffered fundamental shortages: skilled engineers, quality machine tools, raw materials and especially fuel supplies. The German General Staff suffered no illusions about the disadvantages of a two-front war of attrition with their forces on the defensive. The demands of the Eastern Front shielded Britain from all but tip-and-run bombing raids, while American factories (despite their camouflage netting) were totally out of reach. What in Germany was a desperate situation of undersized factories producing outmoded aircraft with a workforce increasingly dependent upon unwilling foreign and slave laborers was in the United States recovery from the Great Depression.

As it turned out, the countries with the greatest resources made the most intelligent production decisions. America mass-produced only 18 basic models. Germany and Japan frittered away resources by building many aircraft types. Japan introduced 90 different warplanes of more than 160 variations; Germany concentrated on high production runs of obsolescent aircraft types. Thus, Heinkel He 111s that could not live in the air after 1943 continued in production until the end of 1944.

Apparently it was considered more advantageous to have something in the air to be shot down than to have nothing in the air at all. The Messerschmitt Bf 109 was built until the last day of the war, and even the antique Junkers Ju 52 "Auntie Ju" remained in production until the middle of 1944.

Then there was the matter of student pilots. Britain and the United States established new schools until, like measles, they dotted those areas of the United States and Canada where weather conditions were ideal for flight training. Rotated home from operational tours over enemy territory, combat pilots taught tactics to fledgling aviators. Flight students received hundreds of hours of experience, including work in formation and instrument flying, navigation and gunnery. The training effort was so great that in 1944 surplus flying classes began to be sent to infantry school in expectation of the planned invasion of Japan.

In contrast, German flying schools lacked everything but brave student pilots. Limited fuel, nonstandardized aircraft (often a mixed bag of Czech, German, French and even captured American types) and a dwindling supply of instructors hurt the training effort. Teachers themselves were repeatedly tapped to fly operational sorties. Later in the war a German flying field became a victory buffet for passing P-51 pilots. The Japanese were little better off, and the results of decreased training became apparent in the battles of 1944.

Once again it was a matter of scale; the Germans and Japanese seemed almost offended that the size of armies and air forces that they could field was not adopted as standard by the Allies. Instead, the incorrigible Allies began 1,000-plane raids, day and night bombing and fighter escort all the way to the target and return. They employed thousands of strategic bombers, those beastly four-engined B-17s, B-24s, Lancasters and B-29s. Against nine or ten Messerschmitts in a patrol, the Allies put up clouds of P-38s, P-51s, Spitfires and Yaks. At this point truth must have come home to Axis leaders; not only is war hell, it is not fair!

Of course, the Allies also had their problems, more awk-

Below, in this artist's depiction, daring Avro Lancaster crews of the RAF's 617 Squadron skipped special bombs into the Ruhr Valley's Möhne Dam on the night of May 16, 1943, flooding part of Germany's industrial heartland.

ward than terminal. British night bombing was at first abysmally bad, with navigation so inaccurate that, for instance, fewer than one bomber in ten came within *five miles* of its assigned target in the haze-shrouded industrial Ruhr. In a series of raids on Cologne, the British dropped 6,600 high-explosive bombs and 147,000 incendiaries, but according to methodical German records only 1,100 of the former and 12,000 of the latter hit the target. The cost of this low "yield" was going up; there was an attrition rate of 1.6 percent of all sorties during 1940 for the RAF Bomber Command. By late 1941, the rate was 4.8 percent and rising.

Churchill, assailed at home and abroad to open a second front, was determined to press on with his only offensive weapon; more and more resources were devoted to the creation of the "4,000 bomber force" demanded by Air Chief Marshal Sir Arthur "Bomber" Harris. "Area bombing," a euphemism for indiscriminate dumping of bombs in urban centers, came about by default. A series of inventions made bombing more precise. The first was Gee—a radio aid to navigation—and later, Oboe, a ground-controlled device to guide bombers to the target. Britain's fastest and most maneuverable light bomber, the de Havilland Mosquito, was assembled into groups responsible only for locating and marking targets, often at a very low level. For such "suicide" missions, highly motivated volunteers received tough, specialized training.

By scraping up every bomber able to fly, Harris sent 1,064 aircraft against Cologne, the first of the "millennium" raids in late May. In July 1942, the first Boeing B-17 of the U.S. Army Eighth Air Force, the harbinger of future air might, landed in Prestwick, Scotland.

Axis innovations included the first resupply by air of surrounded ground forces—a prelude to Stalingrad, when the Germans maintained troops by large-scale glider forces at Demyansk and Kholm, south of Leningrad. The Germans handed British ego a stunning blow when famed German

B-17 Flying Fortresses, left, of the 96th Bomb Group, Eighth Air Force, based in England, press on through large-caliber flak over northern Germany, 1944. This Eighth Air Force shoulder patch, right, was worn by Major Reynolds Benson, above at center stage, briefing 96th Bomb Group crews at Poltava, in the Soviet Union, June 26, 1944. During this first "shuttle mission," June 21–July 5, 1944, Eighth Air Force bombers from 10 groups flew from Britain to the Soviet Union, then to Italy and back to Britain, bombing Axis targets on each leg.

fighter pilot and leader Adolf Galland masterminded aerial coverage of the battleships *Gneisenau* and *Scharnhorst* and heavy cruiser *Prinz Eugen* on their dash through the English Channel. They had slipped the relative safety of the French port of Brest on their way to haunt the cloudy coasts and fjords of Norway. The Japanese made a powder-puff retaliation for Doolittle's Tokyo raid, sending a Yokosuka "Glen" submarine-launched reconnaissance aircraft, which managed, almost unnoticed, to drop only four incendiary bombs into the forests of Oregon.

The Grumman F6F Hellcat and the Boeing B-29 made their first flights in 1942. These planes would figure prominently in the defeat of Japan. A supporting role would also be found for the new Northrop P-61 Black Widow, a radar-equipped night fighter. Germany, which again seemed to be just half a war too late, introduced the Messerschmitt Me 262, the world's first operational jet fighter, and the four-engine Me 264 prototype, lackluster counterpart to the B-29, flew and was called with optimism, the *Amerika Bomber*. More ominously than we knew, the scientists and slaves at Peenemünde fashioned the A-4 (V-2 when deployed), the first rocket-propelled ballistic missile of international capability, later launched against London and Antwerp.

By 1943, the Allies began to hammer out air superiority in the skies over Germany. No major campaign would ever go well again for the Axis powers: 1943 would see Germany in full retreat in Russia, thrown out of Africa, deserted by Italy and harassed night and day by air. Island by island, Japan would be knocked back from the Solomons toward Rabaul, as her maritime lifeline was mercilessly bombed, torpedoed and gunned out of the water.

A sad and almost inexplicable warfare evolved, in which the veteran brave and resourceful German and Japanese troops fought with skill and daring—almost without air power—giving their lives for hopeless causes.

A small band of villains at the head of each country, men

Home-front support for the immense overseas war effort included movies and songs by well-known performers. Some entertainers, such as B-17 pilot James Stewart, flew combat missions; many others made arduous flights on military air transports to entertain war-weary troops all over the world.

who should have known better, who must have realized that the war was long lost, kept sending out millions of their countrymen to die in order to prolong their own existence by two or three years. When the Allies presented their overwhelming power in the Mediterranean, the Italians gratefully rid themselves of Mussolini and his Fascist straw facade and surrendered. The Germans and the Japanese could have displayed the same realism at the same time, and perhaps ten million more people would have survived.

Though Italy kicked out Mussolini and surrendered in 1943, the Allies did not expect such sanity from Germany and Japan. The relentless buildup of the overwhelming Allied air forces continued. Rare metals for aircraft alloys and good lubricants became increasingly scarce for both Germany and Japan. Japanese quality control fell off so badly that the development of more powerful engines like the Kasei 23a for the Mitsubishi J2M "Jack," or the Kawasaki Ha-40 for the sleek "Tony" were dogged by problems.

Both Germany and Japan created stopgap solutions, both demanded more willpower, more self-sacrifice from their aircrews and from the ground and naval forces. But the rest of the air power equation was being rubbed from their boards. Training was reduced and reduced again: pilots were sent into combat with barely 100 hours flying time in trainers and with none at all in the operational aircraft they were required to take into battle.

For the first time Allied aerial bombardment was beginning to build to levels that would begin to achieve what Italian General Giulio Douhet and other interwar prophets of air warfare had forecast for future conflicts. U.S. Major General Ira Eaker and British Air Chief Marshal Sir Arthur Harris initiated the Combined Bomber Offensive Plan—Operation Pointblank—which would optimize the joint effort against the Nazis. The United States Eighth Air Force was to attack airframe and aircraft component factories by day, while by night Bomber Command was to pound urban areas that contained a preponderance of industry oriented to aircraft manufacture. It was a one-two punch, a jab and a roundhouse right.

The Germans reacted forcibly, drawing planes and anti-aircraft guns from the Russian front in an attempt to fashion a roof over *Festung Europa*; properly defended, Europe could indeed be a mighty fortress. Armed with four 20mm cannon, Focke-Wulf FW 190s were able to gun down bombers, and even the older Messerschmitt Bf 109s and Bf 110s were effective when the Allied intruders flew beyond the range of escort fighters. On one of the first raids against one of the so-called panacea targets, the ball-bearing plants at Schweinfurt, 60 out of 288 Allied bombers failed to return. This attrition rate—more than 20 percent—was not only unacceptable; it jeopardized the whole concept of precision daylight bombing.

Substantially improved for controlling night fighters, German radar guided Bf 110s, Heinkel He 219s, and Junkers Ju 88s and others to intercept British bombers. The attrition rate soared past five percent, affording each crewman the sure knowledge that he could expect to die in 20 missions or less, at least before the introduction of Window—the shredded tinfoil antiradar device successfully used over Hamburg during massive raids that incinerated much of this great port city and killed thousands of its inhabitants. Other new tactics helped British Bomber Command improve the odds for the survival of their aerial attack forces.

Yet in terms of either interrupting production or destroying morale, around-the-clock bombing failed. Production soared for critical elements and morale stayed high even where as much as 70 percent of the housing in some cities was destroyed. Part of this stemmed from the fact that Germany had not truly mobilized before late 1943, and part from the genius of Speer for concentrating on essentials.

Standard bomb load for this Eighth Air Force B-17G was 6,000 pounds, and the aircraft carried up to 13 .50 caliber machine guns for defense. Overleaf: P-51D Mustangs of the Confederate Air Force, a Texas organization that restores and flies World War II aircraft. With its superb maneuverability, 437-mile-per-hour top speed and range unequaled by any other single-engine, land-based fighter, the P-51D is considered by many to have been the best fighter of World War II.

GUNFIGHTER II
AAF SPEC. PROJ NO. 39015R
U.S. ARMY P-51D-25-NA
SERIAL NO. AAF 44-73264

G
REGIS F.A. URSCHLER
U

Thus it can be argued that in many respects the air war through the first half of 1944 was merely a continuation of the kind of air war that started in 1914. To be sure, operations were carried out with greater effect and over longer distances and at higher speed; but the result was no more definitive. Hardy soldiers could still hole up in the rubble and, supported by a few tanks, hold off hordes of attackers—at least for a while. Laborers still arose from cots stretched out in schools and subways and tottered off for another 16-hour day in a factory with no roof but still equipped with fine machine tools. Railroads and barges ran at their peril, but they ran, and the gigantic German subcontracting system could still produce enough weapons to arm its own troops and those of the remaining satellite armies.

Japan evacuated the tropical island of Guadalcanal in 1943, just as they were forced to do in the arctic Aleutians. Carriers were streaming from U.S. shipyards in numbers the Japanese could not even comprehend. And as an insolent fillip to the air war, Lockheed P-38G Lightnings made a brilliant flight of 550 miles to the split-second interception of Admiral Isoroku Yamamoto. One of the P-38 pilots, Thomas Lanphier, blew the wing off Yamamoto's "Betty" bomber. The mission was a symbol of the unremitting punishment being dealt to the Japanese and a warning that the homeland would soon be exposed to suffering on Germany's scale. As with the Battle of Midway in 1942, the aerial intercept of Japan's greatest naval hero had been organized around facts gained from deciphered Japanese radio messages.

"Easy Aces" might title operations over the Mediterranean. Here Allied forces attacked formations of the gigantic six-engine converted glider, the Messerschmitt Me 323s, usually escorted by hapless Bf 110s. Big airgoing transports, the Gigants carried fuel, supplies and wounded troops, and were perhaps the easiest targets of the war.

During May 1943, Wing Commander Guy Gibson, V.C., led his No. 617 "Dam Buster" squadron to skip Barnes Wallis's special "earthquake" bombs up against the Möhne, Eder and Sorpe dams in Germany's industrial Ruhr. In the "Med," the little island of Pantelleria yielded to air power after an intensive bombing. Then, in an encore, the Italian commander of nearby Lampedusa surrendered to British Sgt. S. Cohen who had force landed his Swordfish torpedo plane when he ran out of fuel. As a bonus, he accepted the surrender of two other small islands as well, even though he was unable to force land on them.

An endless supply of planes, bombs, crew members, guns and ammunition arrived in Great Britain as the island kingdom was gradually converted into the most versatile and quite unsinkable aircraft carrier in existence.

The USAAF made its deepest bombing penetration to date on August 1, in a low-level attack against the heavily defended Ploesti oil fields in German-occupied Rumania. Out of 174 aircraft dispatched, 54 were lost, with only 33 flyable after the raid.

Italy surrendered on September 8, sending the remains of her fleet south to Allied territory. The Germans were not pleased by the desertion of their ally. The Luftwaffe attacked the Italian battleship *Roma*, sinking her with Fritz X missiles launched by Dornier Do 217 bombers. She was the first—and one hopes the only—battleship to be sunk by a cruise missile. The Germans further restored the Luftwaffe's luster by performing an airborne rescue of the deposed and imprisoned Mussolini, snatching him off of Gran Sasso Mountain in a Fiesler Storch and lifting him to a new career as *Il Duce* of the Salo Republic.

In America, a new generation of transport aircraft took to the air with the flights of the Douglas C-54 and the Lockheed C-69 Constellation. Germany flew the world's first operational jet bomber, the Arado Ar 234. The British launched two jet fighters—the Gloster Meteor and the de Havilland Vampire, designs that would serve them well in the future, if not in this war.

But the real air power news was the buildup of the Boeing B-29 Superfortress effort. The giant Boeing was second only to the Manhattan Project in terms of investment and importance—all of the problems attendant to manufacture, service, training and deployment were being solved. History would show that when they were solved, Japan would be ravaged. They were solved in 1944. In the end, the Manhattan Project was to lead to the creation of the nuclear bomb, and the Superfortress was to deliver the first two—and one hopes the only—such weapons used in anger.

During 1944 the United States produced nearly 100,000 of the total of 167,654 aircraft built by the Allies. With frenzied effort, Germany somehow produced 39,807 aircraft at a time when it was being bombed into rubble. And Japan reached its

As the Nazi Third Reich tottered toward collapse under massive assaults from the air and the great Allied armies closing in from East and West, artist Harold Schmidt pictured USAAF Martin B-26 Marauders adding to the Nazi's pyre as ancient gods battled in the Götterdämmerung*—the "Twilight of the Gods" of Germanic mythology.*

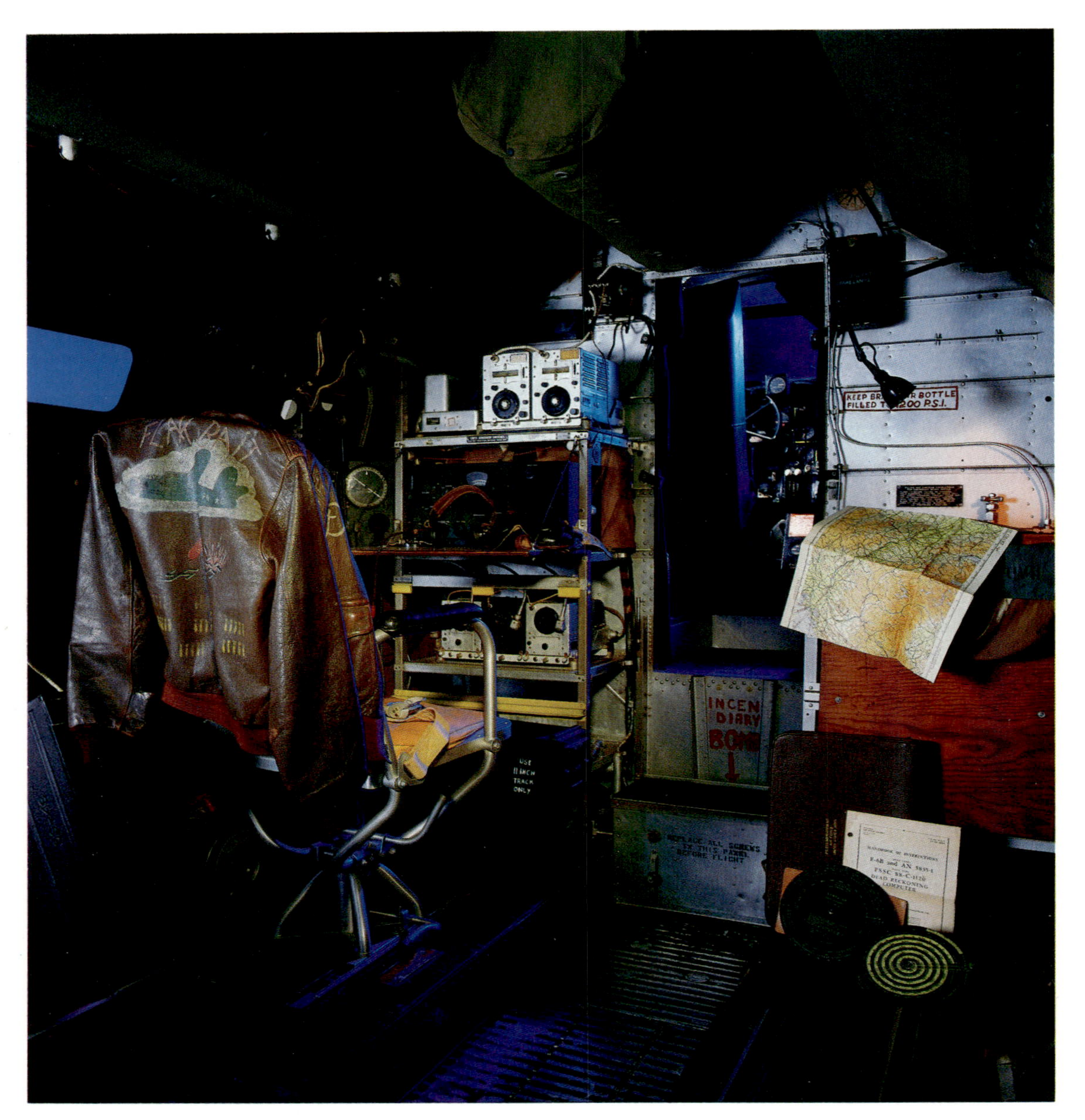

Forward section of Flak Bait, *a famed Martin B-26 bomber, resides in the World War II hall of the National Air and Space Museum. Here, looking toward the cockpit, we see the radio operator's niche. Legendary for its invincibility,* Flak Bait *completed a record number of missions over Europe, in excess of 200. More than 1,000 patches in its fuselage hide perforations from shell fragments—called flak from the German word for antiaircraft cannon,* Fliegerabwehrkanone.

From the detailed mural painted by Keith Ferris at the National Air and Space Museum: B-17s of the 303rd Bomb Group are jumped by German fighters as they leave the target area over Wiesbaden, Germany, in August 1944. Colorful nose art was common on American aircraft as were the row of bombs marking missions on veteran B-17G Thunder Bird.

1920

...ght R.B. Racer

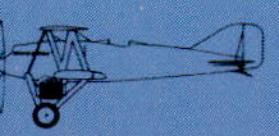
Curtiss CR-1

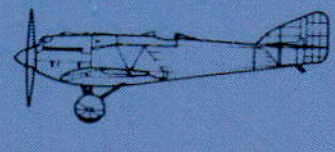
Verville-Sperry R-3

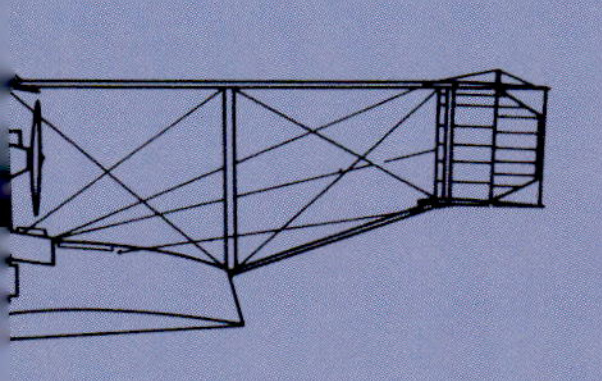

Waco 9

Curtiss P-1 Hawk

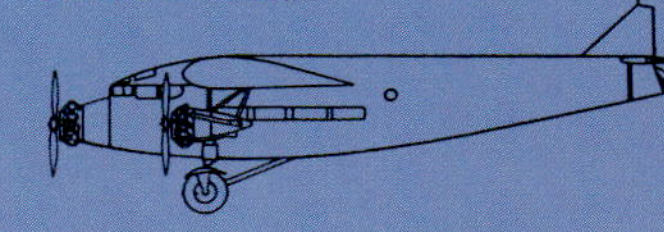
Ford 4-AT Tri-motor

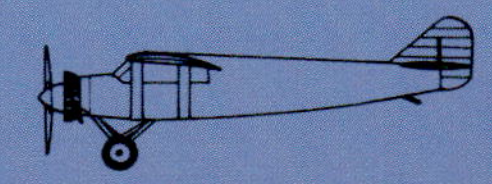
Stinson SM-1 Detroiter

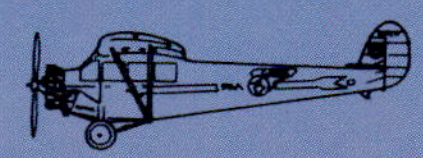
Fairchild FC-2

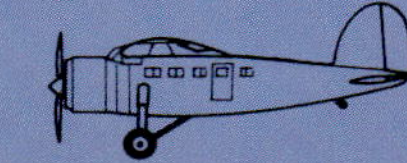
Lockheed Vega

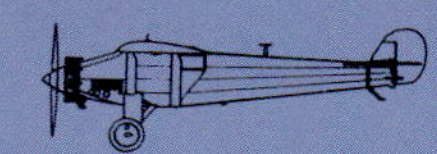
Ryan NYP *The Spirit of St. Louis*

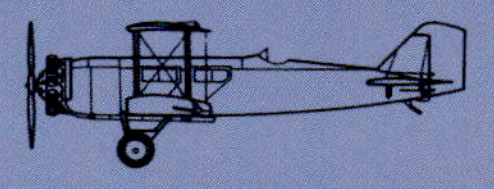
Boeing Model 40

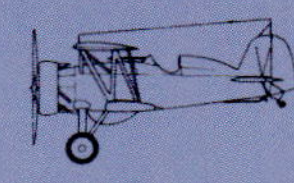
Boeing F4B

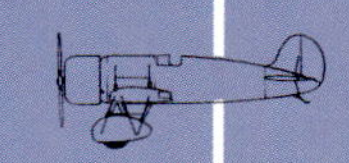
Travel Air Model R *Mystery Ship*

1930

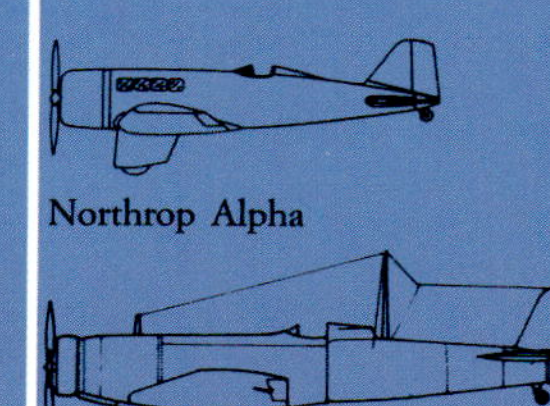
Northrop Alpha

Boeing Model 200 Monomail

Beech Model 17 "Stagger Wing"

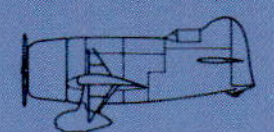
Gee Bee R-1 Super Sportster

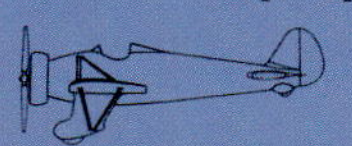
Boeing P-26 "Pea-Shooter"

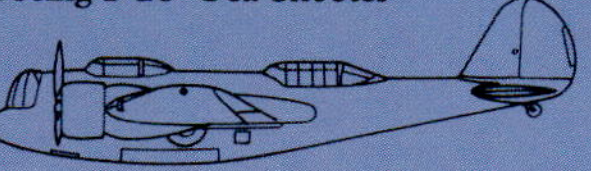
Martin B-10

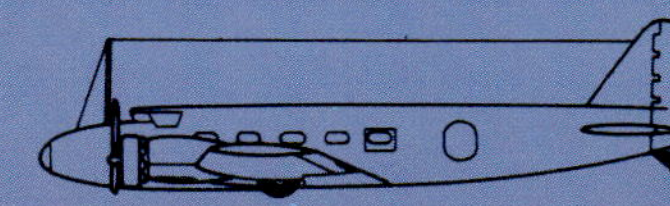
Boeing Model 247

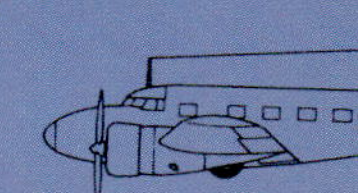
Lockheed Model 10

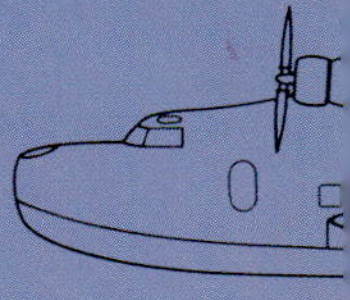
Martin M-130 *China*

C
238050
U
THUNDER
BIRD

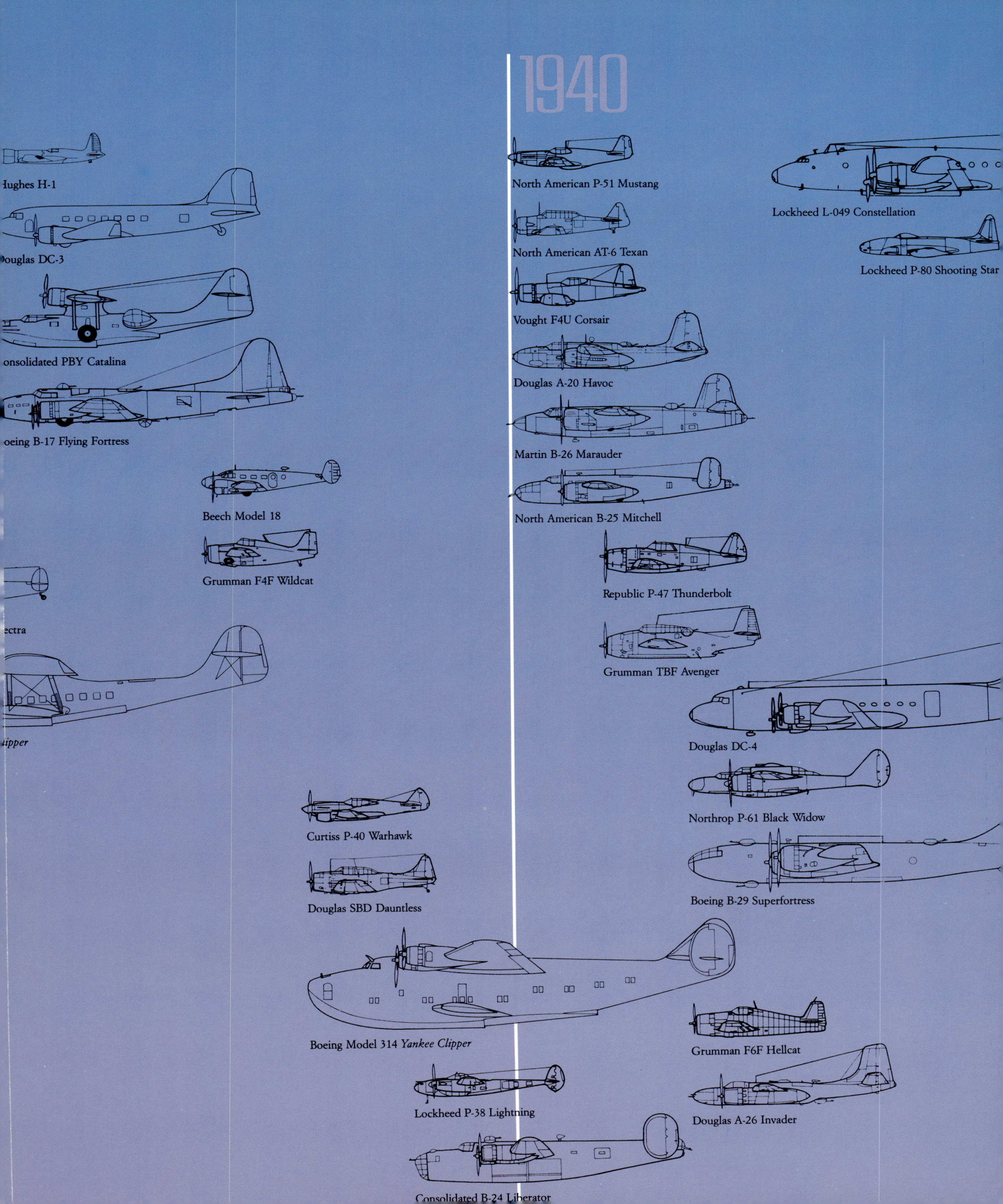

1940
Hughes H-1
Douglas DC-3
Consolidated PBY Catalina
Boeing B-17 Flying Fortress
Beech Model 18
Grumman F4F Wildcat
ectra
ipper
Curtiss P-40 Warhawk
Douglas SBD Dauntless
Boeing Model 314 Yankee Clipper
Lockheed P-38 Lightning
Consolidated B-24 Liberator
North American P-51 Mustang
North American AT-6 Texan
Vought F4U Corsair
Douglas A-20 Havoc
Martin B-26 Marauder
North American B-25 Mitchell
Republic P-47 Thunderbolt
Grumman TBF Avenger
Douglas DC-4
Northrop P-61 Black Widow
Boeing B-29 Superfortress
Grumman F6F Hellcat
Douglas A-26 Invader
Lockheed L-049 Constellation
Lockheed P-80 Shooting Star

American Aviation Chronology

These 110 aircraft provide a representative selection of some of the best or most important United States aircraft from each decade. The group is not intended to be all-inclusive. The nose of each aircraft marks the approximate first flight date of the prototype, unless a particular model is specified. When an aircraft, such as the Douglas DC-7C, has changed dramatically from its original prototype, the date indicates the first flight of that particular model. Italics reveal names given individual aircraft, while those in quotes are nicknames for types, names more popularly recognized than the manufacturer's designation.

1900

Wright Flyer No. 1

Wright Flyer No. 3

Wright Flyer Model A

AEA Aerodrome No. 3 *June Bug*

Curtiss-Herring No. 1 *Reims Racer*

Wright Military Flyer

1910

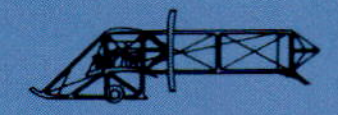
Wright Modified EX *Vin Fizz*

Curtiss A-1 *Triad*

Martin TT

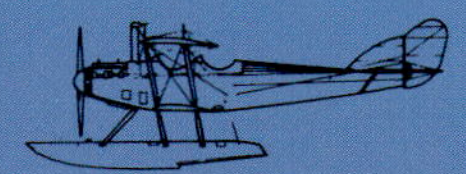
Curtiss N-9 Seaplane

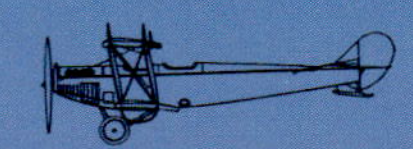
Curtiss JN-4 "Jenny"

Thomas Morse S-4

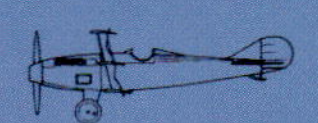
Curtiss S-3

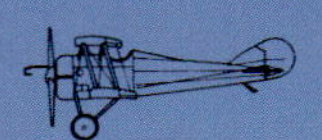
Standard E-1

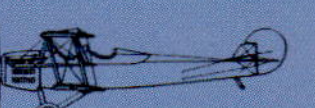
Lewis & Vought VE-7

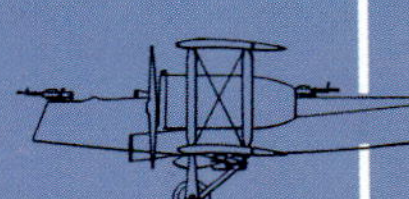
Martin GMB Bomber

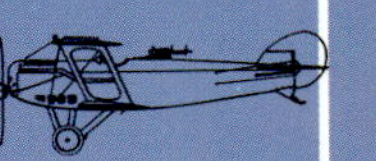
Packard LePere-LUSAC II

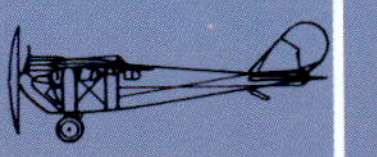
Loening M-8

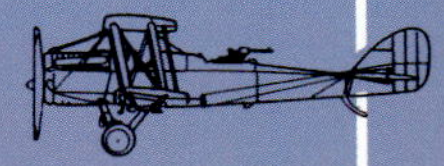
Dayton Wright DH-4

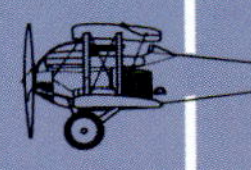
Thomas Morse M

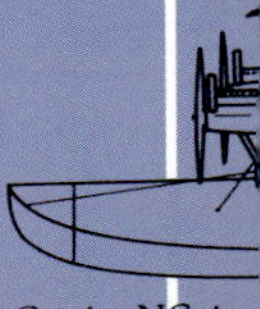
Curtiss NC-4

19

Dayton

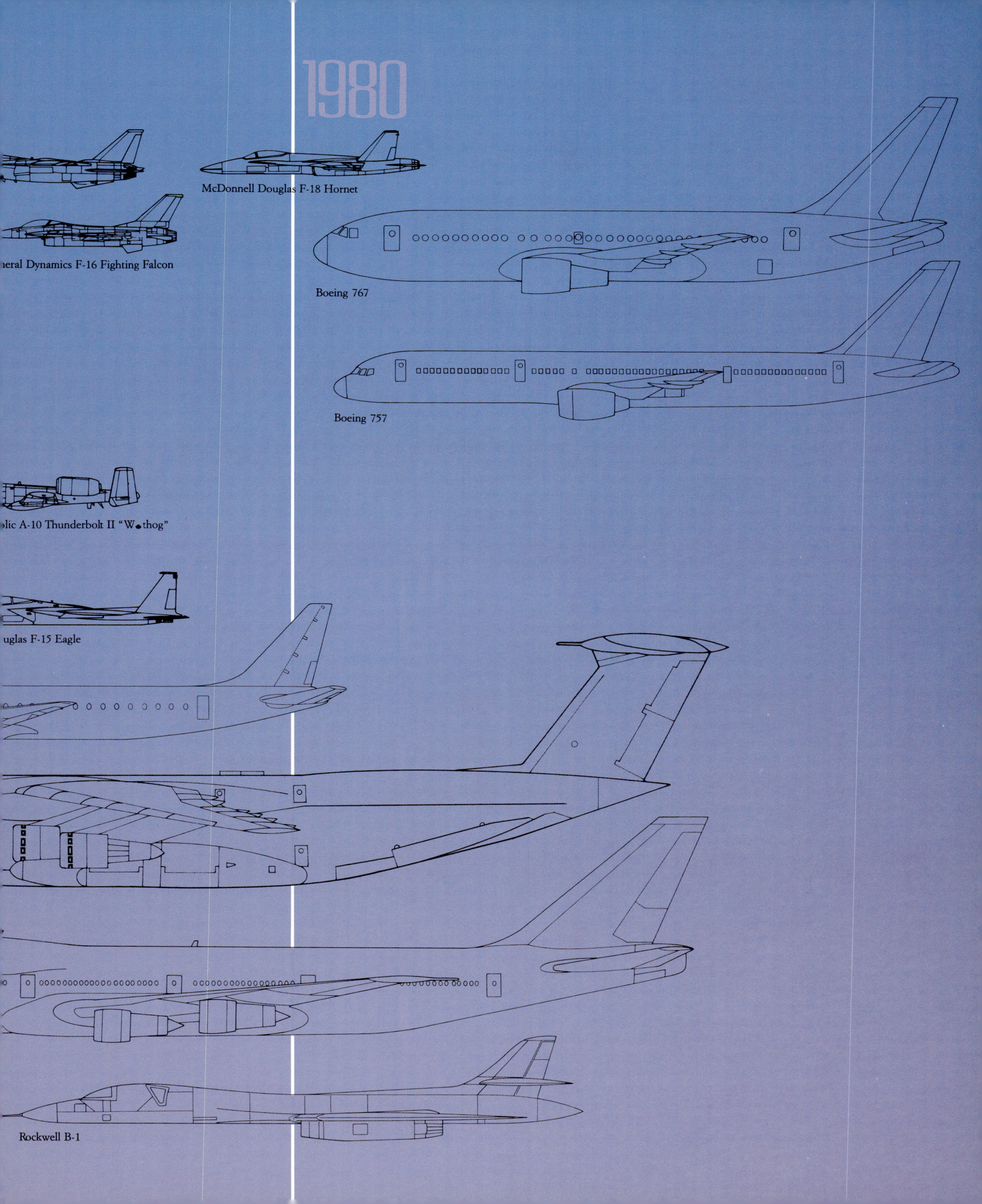
1980
McDonnell Douglas F-18 Hornet
neral Dynamics F-16 Fighting Falcon
Boeing 767
Boeing 757
blic A-10 Thunderbolt II "W thog"
uglas F-15 Eagle
Rockwell B-1

METZ
AEF

1960
1970
Grumman A-6 Intruder
Lockheed SR-71 Blackbird
Boeing 727
Gates Learjet
McDonnell F-4 Phantom II
Lockheed C-141 Starlifter
North American X-15
General Dynamics F-111 "Aardvark"
Douglas DC-9
Douglas DC-8 "Series 60"
Lockheed C-5 Galaxy
Boeing 747

Give me the man who
with a song in his
Popular Songs of the A.E.
Y.M.C.A.
ARMY
HOUSEWIFE
FIRST-AID
LEAVE OR DUTY RATION BOOK
SOLDIER OR SAILOR
COMPUTER
TYPE D-4
PROPERTY OF U.S. ARMY AIR FORCES

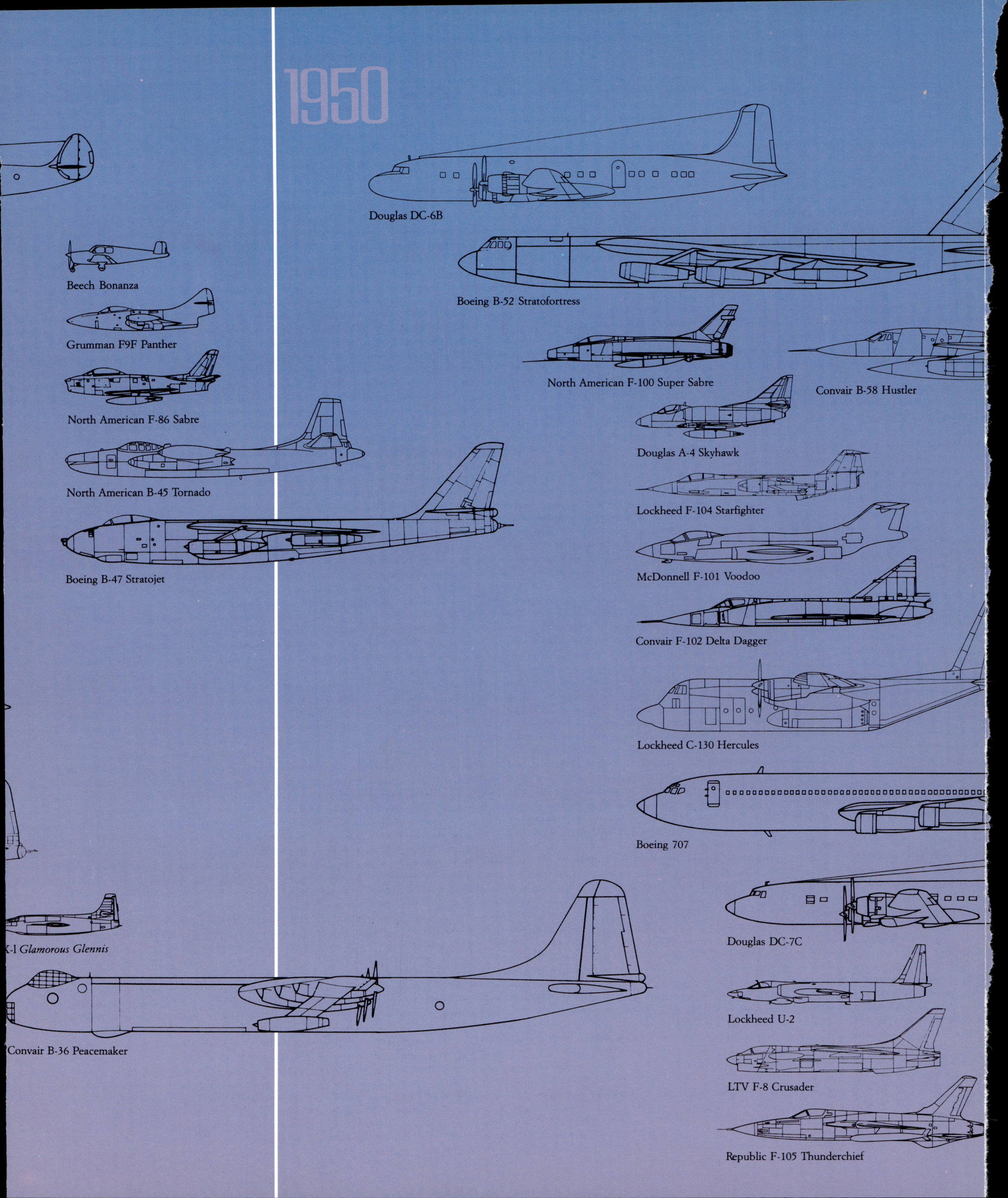
1950
Douglas DC-6B
Beech Bonanza
Grumman F9F Panther
North American F-86 Sabre
North American B-45 Tornado
Boeing B-47 Stratojet
X-1 Glamorous Glennis
Convair B-36 Peacemaker
Boeing B-52 Stratofortress
North American F-100 Super Sabre
Convair B-58 Hustler
Douglas A-4 Skyhawk
Lockheed F-104 Starfighter
McDonnell F-101 Voodoo
Convair F-102 Delta Dagger
Lockheed C-130 Hercules
Boeing 707
Douglas DC-7C
Lockheed U-2
LTV F-8 Crusader
Republic F-105 Thunderchief

AVIATION CADET
TRAINING
FOR THE
RESTRICTED
Survival at
ALTITUDE
of HEAVY
BOMBER CREWS
DOCTRINATION UNIT
Я американец
"Ya Amerikanets" (Pronounced as spelt)
Пожалуйста сообщите
сведения обо мне в
Американскую Военную
Миссию в Москве
Please
communicate
my particulars
to American
Military
Mission
Moscow.
THIS IS TO CERTIFY THAT
Littleton Pendleton
NAME
RATING
GRADE
SERIAL NO.
ORGANIZATION
SERVED HONORABLY AND WELL IN THE
UNITED STATES ARMY AIR FORCES IN
WORLD WAR II.
COMMANDING GENERAL
ARMY AIR FORCES

Memories of War in the Air

Objects at left are from Smithsonian and private collections.

Unsung heroes of fleet air operations, World War II aviation machinist's mates swarm over a Douglas SBD Dauntless in this painting by James Dietz. Aircraft needing involved repair or service were transported on giant elevators from the bustling flight deck to the cavernous hangar deck with its machine shops.

absolute potential with the manufacture of 28,180 planes in a single year. In the United States, boards were formed to determine the kinds of aircraft to cut back and ways to dispose of the reserves. At the end the unopposed Allied air patrols attacked new German airplanes as they were towed or pushed from the assembly line to storage. The aircraft that survived would never fly because of the lack of fuel. Japan had yet a little respite before the B-29 raids became effective, but was chronically short of fuel and trained pilots.

In late 1943, the USAAF began long-range operations with fighter escort; it spelled not only the death of the Luftwaffe, but the piecemeal destruction of even the minor sinews of the German supply system. The German fighter force was lured into battle, smashed, harried to the ground, then pulverized on the airfields and in the factories.

The German pilots were no less brave than they had been earlier, but most were less skillful. The hard core remained—the incredibly experienced old stagers, the Steinhoffs, Ralls, Barkhorns, Nowotnys, Krupinskis—but even the skill of these men in their new jet fighters could not stave off the hordes of Tempests, Typhoons, Mustangs and Thunderbolts. The 1944 Normandy invasion provided the first example of how Allied air supremacy could be used: the Luftwaffe was simply brushed aside.

Then, as the Allies established forward airfields in France, life became even more intolerable for the German defenders. The Allied fighters met no resistance and ranged free, shooting up trains, barges, boats, trucks, bicycles and carts—anything that moved. And only then did the brickwork of the German economy really begin to fail, even as the individual bits of human mortar were shot away. After mid-1944, British Bomber Command and the United States Eighth Air Force combined their capabilities and began to smash Germany in earnest. This was a new kind of air power, very different from all that had been deployed before. The invasion confirmed on the ground what had been written in the sky. The East was not a front; it was the Apocalypse. Every effort turned against the West brought the Russians nearer the German frontier.

For the Japanese, 1944 opened with the fall of Kwajalien Island, moving the Allies 1,000 miles closer to Tokyo. General Douglas MacArthur returned as he had promised to the Philippines, where the Japanese fleet made its last desperate sortie to fight without air cover, relying on the new Special Attack Corps—the first of the Kamikaze attacks, a futile if gallant gesture. A number of Allied vessels were sunk or

At the time of its introduction in 1944, the Boeing B-29 bomber, below, represented perhaps the most complex engineering achievement in history. Its speed and range were unsurpassed by any other heavy bomber in World War II. Left, B-29 Lucky Strike *was pictured during a spring 1945 raid over Tokyo by artist William S. Phillips in honor of General Curtiss E. LeMay, commander of the B-29 fleets that all but destroyed many Japanese cities in the last year of World War II.*

damaged, and many American sailors died or were fearfully injured in the Kamikaze attacks, but U.S. manpower and industrial might could absorb such losses. Only in China did the Japanese seem to have any strength, and they made a telling effort to capture the American air bases in the south, where B-29s were beginning their systematic destruction of resources in the Japanese home islands.

The crucial outer islands fell one by one, forcing the Japanese defensive perimeter ever backward. As each island fell, two "flames" were brought closer to Japan. The first was that of the B-29s' incendiaries; Japan's cities of wood and paper provided ideal targets for air warfare brought to maturity—they burned well. The second was the nuclear flame, for which all targets were perfect.

In Europe, the concept of precision bombing became a reality, symbolized by a flight of Mosquito bombers that hit a German prison at Amiens, France, to liberate resistance patriots awaiting execution. Already, strategically speaking, the noose was tightening around the exposed neck of the German homeland.

At the dawn of 1945, Luftwaffe chief Göring had roused his men for one last strike—Operation *Bodenplatte*. On January 1, they launched a surprise strike against Allied airfields in belated support of the last great German offensive in the West, the Battle of the Bulge. The Germans paid dearly for this audacity, for they could replace neither their lost airplanes nor their pilots.

There was still war in Europe, but only a war of relentless pursuit and execution. Men and women would still die in the ruins of German cities, but no one save perhaps Hitler in his own bunker saw any way out. Bomber operations began to grind to a close as many targets were deemed "no longer profitable" by Allied planners. A grievous exception was Dresden; in a mid-February raid this beautiful, historic and militarily unimportant city was consumed by a fire storm started by bombs, resulting in a terrible loss of life.

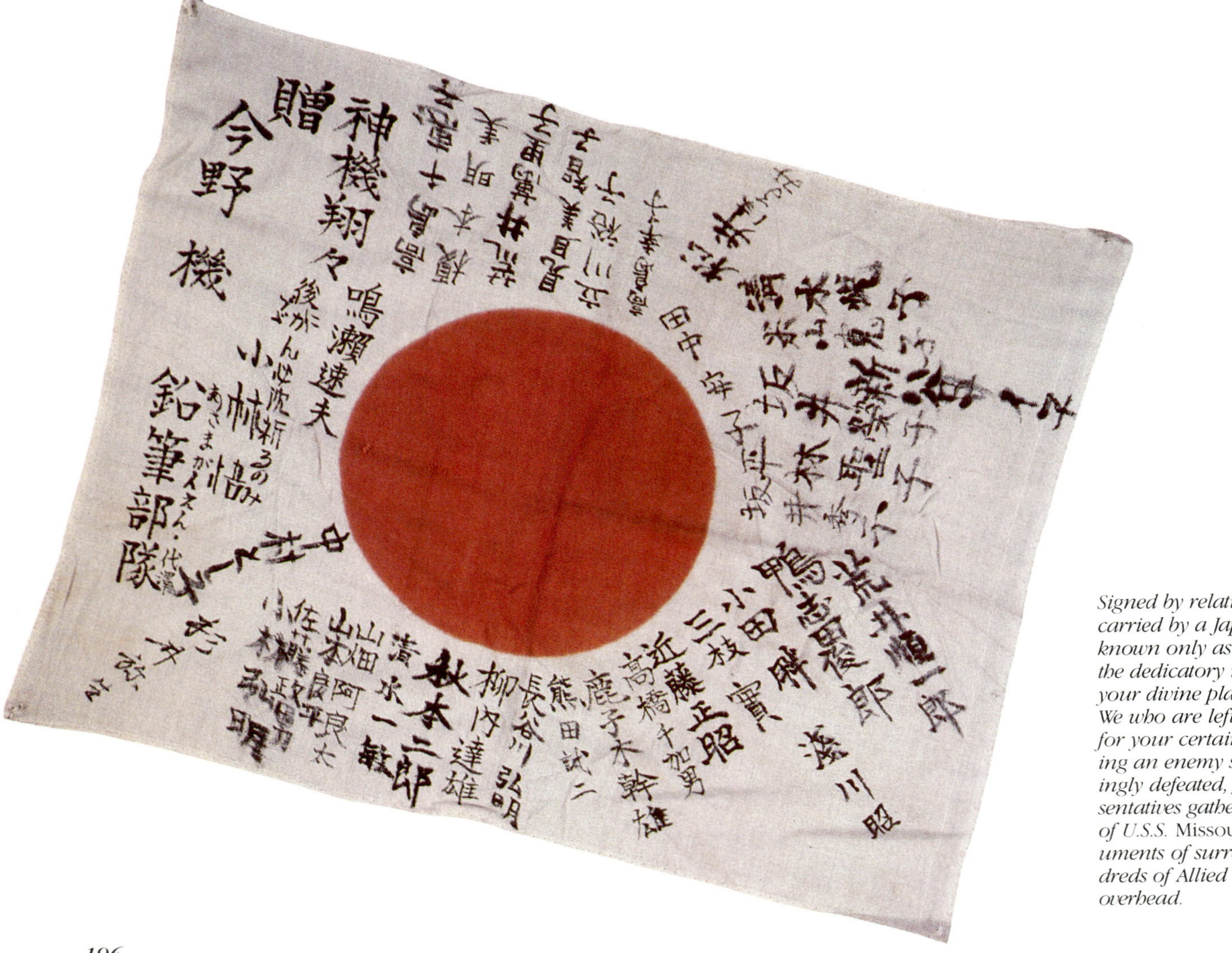

Signed by relatives, a small flag carried by a Japanese pilot known only as Imano contains the dedicatory message, "Let your divine plane soar in the sky. We who are left behind pray only for your certain success in sinking an enemy ship." Overwhelmingly defeated, Japanese representatives gathered on the deck of U.S.S. Missouri *to sign the documents of surrender as hundreds of Allied aircraft passed overhead.*

During March and April 1945, the Allies totally dominated the air, flying where they willed, attacking anything that moved. The Allies put combat air patrols over German airdromes, as pleased to shoot down someone taking off or landing as to best them in a dogfight. A few German planes conducted the last Luftwaffe raid against England on March 20. This brought the total wartime figure to 74,172 tons of bombs dropped by Germany against Great Britain. By May 8, 1945, the British and Americans had dropped a total of 1,996,036 tons (27 times the German tonnage) on Germany, which had sowed the wind and now reaped the whirlwind.

Daylight firebombing from high altitudes during March 1945 shook Japan from one end of its island chain to the other, but the director of the effort, General Curtiss LeMay, was not satisfied with the results. He decided to strip his aircraft of armament and send them on low-level raids at night. On March 9, more than 300 B-29s carrying 2,000 tons of bombs departed from runways on Guam, Saipan and Tinian. More than 280 reached Tokyo and dropped their bombs against moderate resistance; 14 were lost. On the ground a fire storm raged in which 80,000 Japanese perished. Tokyo almost ceased to exist as a habitable city.

The B-29s turned to other cities—Nagoya, Osaka, Kobe and Nagoya again. In ten days, 1,600 sorties had been flown against four major cities. Organized resistance on Iwo Jima ended on March 16, and the B-29s had a safe haven used by more than 2,251 of these giant bombers. Okinawa was next on the list, and the doomed final sortie without air cover of the superbattleship *Yamato* could not save it.

In Japan as in Europe, targets were no longer "profitable;" yet the Japanese fought on. Allied planners turned their eyes toward an invasion. Then World War III in the air began over Hiroshima and Nagasaki. In an instant, two bomber raids made all previous air war almost trivial. In the past, lives, battles, armies, navies and perhaps even countries depended upon the result of air power. Now all humanity did.

PART 3

BY A. SCOTT CROSSFIELD

World War II ended with the largest civil and military air fleet ever. World War II produced the root technology that would render that fleet obsolete. World War II also produced a generation of mature, self-confident and committed young men who would dominate the aerospace arena for the next 30 years. Their philosophical reaction to the years of destruction gave rise to an insatiable appetite for development and building.

In 1947, the Bell X-1 demolished the "sonic barrier," (which was only an institutional barrier) and set the pace for the Research Airplane Program—the cutting edge of aeronautical experimentation since World War II. Operated by NACA (the predecessor of NASA) and funded by the US Air Force and Navy, this program led to safe, casual and low cost mass air travel for all. In addition, the rocket engines that first appeared in the X-1 became the mainstay of hundreds of manned rocket flights on the leading edge of flight for 30 years and the basis of man's venture into space.

During the Fifties, one of the Navy's contributions to the Research Airplane Program was the D-558-2 Skyrocket; it gave us critical experimental information for transonic and supersonic sweptwing commercial and military fleets. A decade later, the USAF's North American X-15 flew to nearly Mach 7 above the atmosphere

Most successful of all experimental research aircraft, the North American X-15, once piloted by Scott Crossfield, now hangs securely in the National Air and Space Museum. Powered by an engine with 60,000 pounds of thrust, the X-15 rocketed to nearly Mach 7 in 1967, a speed no other aircraft has ever approached.

Jets, Rockets and Realism

and paved the way for the conquest of space. By the early Seventies, aerodynamically shaped lifting bodies, culminating in the X-24B, taught us a way to return to earth with finesse.

Today, the Research Airplane concept is alive in the modern day Aerospaceplane and, as of this writing, promises a future of ready access to all of the earth's atmosphere and orbital space. The potential of aerospace industry growth is amazing to behold, truly mindboggling: the *Orient Express,* an airplane to commute from New York to Beijing in two hours, has moved into the sphere of reality.

Meanwhile, in purely atmospheric flight, the feeble net thrust of the World War II jet engines has increased tremendously. The design engineer, for the first time in aviation history, has become power rich and achieves performance without as much concern for aerodynamic drag. However, progress hesitated for many years at Mach 2, the temperature limit of aluminum. Only recently has the promise of the B-70, the SR-71 and X-15 been pursued, stimulated by the resurgence of interest in the Aerospaceplane. They achieved highspeeds through advanced materials.

For various reasons, operational aircraft, both civil and military, have been frozen at Fifties' performance levels for over 30 years. Instead of improving performance, sophistication of aircraft systems became important. For instance, the capabilities of electronic weapon delivery systems changed the technique of modern war significantly. The fighter pilot has gone the way of the knights of old and the cavalry. The "glass cockpit" has become the norm in military and commercial transports. On-board, all weather capability has freed the general aviation fleet from its fair weather limitations.

The American supersonic transport was not built because of a revolt against its environmental impact. While the British/French Concorde was burdened by the same problem, it was built under the condition that supersonic flight be restricted from flight over land. This forecast its demise as a major force in air transport. However, it has found a viable economic niche as a transoceanic carrier.

Private aviation, the wellspring of national aerospace competence, was born of the loving care of aviators for aviation. During the post World War II period, most private aviation companies fell into the hands of lawyers, bankers and regulators who did not understand how to assure its future. Aviation's roots, however, were deep, and the aviators themselves are taking the initiative back through the home-built and ultralight movement. Annual home-built aircraft sales in America have now surpassed manufactured aircraft sales. Today, many home-built craft substantially exceed the performance and utility of the manufactured product and foretell of a major shift in the private aircraft industry. The common man's dream of liberation from earthly constraints may still have hope.

Of course, the epitome of all this is the aviator and his blood brother the astronaut. The national aerospace effort flourishes because it is a human experience that fires the imagination of the nation.

The Wright brothers, unfettered by institutionalism, worked for five years with their own mental and material resources and brought us the miracle at Kitty Hawk. By his own sheer resolve, Charles Lindbergh defied the established rules and brought the Wrights' miracle into the American lifestyle forever.

And now, Jeana Yeager and Dick and Burt Rutan mortgage everything but their dignity, scorn the establishment and the naysayers, work six years to create an ingenious concept, build an "impossible" airplane and execute a brilliant, non-stop flight around the world, on one load of fuel, in nine days of straining the limits of human endurance. There was no prize or promise of riches. It was there to do and it was hard to do. These aviators have again proven in spades that America is replete with vitality. Oh, what a debt the nation owes to these aviators.

And what of our future? I cannot delve into philosophy and prophesy with confidence but I can with conviction parrot Orville Wright, who, when asked to forecast the future of aviation replied: "I cannot answer except to assure you that it will be spectacular."

Insect-eyed, a DC-9 transport awaits its turn for takeoff on a night cargo flight.

BELL

The Jet Age

Of the "Big Three" Allied powers, only the United States remained immune from the general devastation of World War II. The war brought the nation's power—from its farms, industry, people and military establishment—to a level that exceeded any prewar hopes. Nowhere was this more true than in aviation, which America dominated completely. The Germans had taken the lead in jet and rocket engine development and in some aerodynamic features such as the sweptwing, but all of their experience was available for exploitation during the next generation.

The warring nations had created a global system of runways, fuel dumps and navigation and meteorological facilities. They had the trained talent of hundreds of thousands of pilots to use them. But even more important was the revolution in thinking. If you wanted to go to Africa or to India or to cross the Atlantic, you could fly. The end of the era of great ocean liners was at hand. Travel contracted into a matter of hours, not days. Little wonder that this would become the first real flying generation.

Rocket-propelled, the bullet-shaped Bell X-1 first broke the sonic barrier and led to a new generation of postwar experimental planes. Test pilot Charles E. "Chuck" Yeager stands beside Glamorous Glennis, *the X-1 named for his wife that carried him supersonic in 1947; today it hangs in the Milestones of Flight Hall of Smithsonian's National Air and Space Museum in Washington, D.C.*

U.S. AIR FORCE
U S AIR FORCE
U.S. AIR FORCE
U.S. AIR FORCE

C-47s, "mothballed" in aircraft "boneyards" like this one at Davis Monthan Air Force Base in Arizona, above, often reemerged in airline livery. They were the military version of the DC-3. Many of these legendary ships still fly, as do many Beechcraft Beech 18s, left. C-45s to the military, the Beech classic also made an easy transition to civilian life. Produced in many versions from 1936 to 1969, Beech 18s have served primarily as business aircraft.

Naturally enough, aviation manufacturers wanted to keep manufacturing aircraft: World War II had provided the first—and for most, the only—period of prosperity. Meanwhile, the emerging superpowers were determined to project their power abroad and, as much as budgets would allow, to maintain military preparedness. But in reality, the Western armed forces were dissolving like a July hailstone. During this unprecedented demobilization it was only natural for the military contracts to dry up and for factories to close, their skilled work forces irretrievably dispersed. In addition, thousands of surplus aircraft dotted the landscape in great silver rows around the world, from Arizona to Australia. Some were war-weary combat veterans, often decorated with the slightly risqué art beloved by crews. Freshly minted factory models, perhaps sent through an expensive modification program, were flown directly to storage sites and abandoned after only five hours in the air.

Great legends—most of them true—arose from this process, stories of brand new P-38s shoved off docks or P-51s buried for landfill. California stunt pilot and racer Paul Mantz reportedly became "the ninth largest air force in the world" by buying an entire fleet of B-17s, B-24s and others, literally hundreds of airplanes. He reputedly paid the entire bill for them by draining the high octane fuel out of their tanks and selling it. Others bought engines for a pittance, then reaped a fortune from the silver and other precious metals found in them. The last recourse was always the melting pots of the salvage yards, where great airframes were reduced to aluminum ingots, to reappear a year later in house trailers, siding and cookware.

And some aircraft were sold as is; the government tried to control the sale of warplanes, making sure guns and other potentially combat-worthy equipment was removed, but soon workhorses such as the Douglas C-47 and Curtiss C-46 flowed into the expanding civilian economy.

Many military and industrial planners and politicians saw the need to keep up production capability, to advance new and exciting technologies forged in wartime and to develop the next generation of military aircraft. The realities of the Cold War and its hot implications gave them an opportunity within the next few years. The global threat, chillingly enunciated in Winston Churchill's Iron Curtain speech in Fulton, Missouri, brought the arms manufacturers a trickle of contracts that eventually grew into the current, multibillion-dollar Pentagon budget.

Thus emerged the specter of a military-industrial complex, evoked by President Dwight D. Eisenhower as early as 1959.

Yet the buildup of the military-industrial complex began in the desperate catch-up days of 1941, with war raging across Europe, when thousands of aircraft were ordered from the drawing board. So great was the need to create production facilities, with their trained staffs and subcomponent and material networks, that factories often received orders for airplanes that were not wanted—then or later by the U.S. military. Such was the case with the Vultee A-31 and A-35

Vengeance aircraft, for example. They were built, pure and simple, to achieve and maintain a production capability and to supply our allies. Factories even sprang up for unproven types; one of these, the Curtiss C-76 Caravan, was possibly the worst aircraft of the war. When the flight tests proved unpromising (or life-threatening, as with the C-76) the factories were converted to other work.

Postwar planners had their own style; they placed orders for smaller quantities. The armed services scattered contracts for prototypes, conducting extensive and often competitive fly-offs before choosing a manufacturer and signing the production contracts.

Business opportunities appeared unlimited for the manufacture of the new jet engines. Allison, General Electric, Lycoming, Westinghouse and others felt they could compete with the giant Curtiss-Wright and Pratt & Whitney corporations. Marquardt had its special niche, the ramjet, a device that came into play only after the aircraft reached high speeds. Initially, Pratt & Whitney lagged in development of the jet engine, largely because the authorities had directed them to concentrate on their enormous production lines for piston engines. But Pratt & Whitney soon caught up.

Before the military could plunge into purely postwar designs, they had to deal with the airplanes designed and built during the last years of the war. The gigantic B-36s—destined originally to bomb Germany from U.S. bases if Britain were overrun—became the intercontinental atomic bomber. Perhaps the most beloved misfit in aviation history, the Northrop XB-35 flying wing fluttered into oblivion in the late Forties.

Despite its successes, the Boeing B-29 was underpowered and beset with engine fires. Refitted with the more powerful and ultimately reliable Pratt & Whitney R-4360 "corncob" engines and redesigned from nose to tail, it became the B-50 and ended its days as the KB-50, an aerial kerosene truck for refueling jets in flight. The "Twin Mustang" P-82, a pair of P-51 fuselages joined by wing and horizontal tail sections, had enough performance to warrant manufacture, as did the Lockheed P-80 jet fighter, the first American jet to enter squadron service.

Unlike most other countries during the golden postwar period, Americans seemed to have the time and the luxury to experiment. The first American jet bomber, the XB-43, was clearly an interim aircraft, and the first production jet bomber was the North American B-45, a completely conventional straight-wing bomber—with jet engines. Follow-on fighters to the P-80 Shooting Star appeared, the Republic XP-84 and the North American XP-86, along with follow-on bombers after the B-45, the XB-46, -47 and -48.

In these exciting times, the old U.S. Army Air Force became the U.S. Air Force, a new and separate branch of the armed services. In his presidential airplane, the Douglas C-54 *Sacred Cow,* Harry S Truman signed legislation creating the Air Force on September 17, 1947. That first Air Force flight in the first Air Force One helped set in motion an era of startling technical advancement. At first, though, the Air Force proceeded slowly.

Ordering new aircraft designs, perhaps incorporating untried technology, the military planners usually hedged their bets. Curiously, the Navy's first jet had a propeller. The Ryan Fireball FR-l employed its piston-driven propeller for short takeoffs from carriers, a job later taken over by a steam catapult aboard ship.

While the Republic XP-84 retained a conventional straight wing, the North American XP-86 carried wing and tail surfaces swept back at an angle of 35 degrees. (Doubling back, North American also built the "straight-wing Sabre" for the Navy as the FJ-1 Fury while the FJ-2 was sweptwing after the XP-86 proved successful.) In the bomber competition, Convair was given a piece of cake in the XB-46, a beautiful four-jet airplane that was totally conventional except for the jet power plants.

Martin delivered a more complex aircraft, one with six jet engines and a "Middle River Stump-Jumper" bicycle landing gear. Boeing had the toughest job of all: to combine tandem landing gear, sweptwings and six jets. All these aircraft were successful, but the XP-86 Sabre fighter and XB-47 Stratojet bomber gained immeasurable advantages from their sweptwings and helped place the brand-new United States Air Force well ahead of other countries' air services.

The Air Force really got its money's worth from the B-47. The entire program for two prototypes was set at only $10 million, scarcely enough to field test a new Army napkin today. The newly minted bomber flew on December 17, 1947. Though development was long and sometimes painful, a magnificent aircraft emerged and the X (for experimental) was dropped as it went into full production. Once in operation, the B-47 became the most important military jet aircraft of all time because of its influence on world events and aircraft design. Hot but lovely to fly, the B-47 helped create a "Pax Americana," extending into the Sixties and the era of the missile gap. The Strategic Air Command (SAC) bought more than 2,000 copies of this noble craft.

The North American F-86 Sabre jet fighter ruled the skies over Korea, and in one or another of its many variations was the definitive fighter for most of the air forces of the world for the next decade and a half.

Other nations were not far behind. The Soviet Union surprised the world with its MiG-15 jet fighter, considered by many the equal of the F-86 Sabre in combat if the pilots were equal. Britain generated a series of fighters that began with souped-up Gloster Meteors and de Havilland Vampires and led ultimately to the Hawker Hunter and English Electric Lightning. France pulled itself together after the war, and Marcel Dassault (before the war Marcel Bloch, builder of light, elegant piston-engine fighters) helped to establish his nation in the world arms market with the Ouragan, Mystére and elegant Mirage.

And so history seemed to repeat itself. Once again, the military planners and aeronautical designers began to play their special kind of one-upmanship, as they had since the early days of World War I. Examples of the usual state of affairs—the building of opposing and equal pairs—spring to mind: the Fokker Eindecker against the Nieuport 11, the Albatros versus SPAD VII, Fokker D VII and S.E.5a, Hawker Fury and Curtiss P-6E. Other comparisons can be made: the Gloster Gladiator versus the Fiat C.R.32, the Messerschmitt Bf 109 pitted against the Spitfire, Fw 190D versus P-51, Yak-9 roughly equivalent to the Macchi C.205, F-86 and MiG-15, MiG-21 against the F-4. The Dassault Mirage, Sukhoi 22, IAI Kfir and SAAB Viggen jet fighters share many performance characteristics.

Perhaps because they always design to the "next point" on the performance curve—that one unattainable previously and just achievable now—engineers always seem to be able to produce aircraft of comparable performance regardless of differences in their countries' national resources. Only very rarely do aircraft such as the U-2, the SR-71 or the B-47 come along, aircraft so clearly superior that no other nation can match them. But then, from the Wrights to the great aircraft designers such as Lockheed's Clarence "Kelly" Johnson and Douglas's Edward Heinemann, America always seemed able to come up with surprises.

In postwar America, air travel was available to more and more people. The optimism of the era is reflected in the "aircraft-in-every-garage" theme of a 1951 issue of Popular Mechanics.

Lockheed's elegant Constellation L.749, opposite, served Pan American, TWA and American Overseas Airlines as intercontinental air routes proliferated in peacetime. Pan American opened the first round-the-world service with the "Connie" on June 17, 1947. Competitor to the Constellation, the blunt-nosed Boeing 377 Stratocruiser, top, shared design features with the B-29 and B-50 bombers. Passengers loved this double-decker, with its berths on transatlantic flights and a lower-deck cocktail bar.

This phenomena is not an accident, nor always a product of espionage. It can represent a convergence of technical factors that have become state-of-the-art: the outgrowth of a continuous balancing process. The skillful designer trades off weight against such factors as drag, the amount and type of equipment, maneuverability, pilot safety, comfort, speed, range, takeoff distance, maintenance, reliability and every other element of design. The designer works toward meeting requirements handed down by planners, often from the service or joint-service or even from the defense-ministry level. To reach the goal—perhaps to attain Mach .9 at 45,000 feet—the designer applies the materials at hand and other resources to meet the design specifications. All this happens while the design team strives to retain every other good feature it can to permit fulfillment of their service's operational doctrine. The MiG-15 is a case in point. The Soviet design team of Artem Mikoyan and Mikhail Gurevich was blessed by the unexpected availability of the Rolls-Royce Nene engine design. The sale of this design to the Soviets, a bit of British political expediency, had profound consequences. The Soviets placed high priority on antiaircraft defense especially after witnessing the devastation wreaked upon Japan by atom bombs. Soviet intelligence presumably knew that the infant USAF had, in routine precautionary planning, already earmarked 133 nuclear bombs against 70 major Soviet population centers by 1948. Defense requiring interceptors that had low weight and high punching power was therefore imperative. The result was the MiG-15—a spartan aircraft—weighing in at 12,556 pounds; it carried one 37 mm and two 23 mm cannon and was easily serviced at the then-primitive Soviet system of airfields. In this instance, the Soviet design doctrine closely paralleled that of Douglas designer, Edward Heinemann, who advocated, "Simplificate and add lightness." Lockheed's great designer Clarence L. "Kelly" Johnson provided the inspired acronym "KISS" for "Keep It Simple, Stupid."

The F-86 Sabre, designed to be an air superiority fighter, was very sophisticated, with its tail inspired by the Bell X-1 and leading-edge slats for low-speed flight courtesy of the Messerschmitt Me 262. Other features included large speed brakes and hydraulically boosted flight controls. With a gross weight just shy of 16,000 pounds, its ceiling was 48,000 feet and its top speed 679 miles per hour. On balance, the MiG-15's ceiling was rated 4,000 feet higher and its top speed 11 miles per hour lower than the F-86.

TWA
TRANS WORLD AIRLINE

Brainchild of John K. Northrop, the YB-49 Flying Wing came close to low drag perfection. The promising Flying Wing project was dropped from the military's program in 1950. Unofficial reports ascribe a flying-wing design to the highly classified "Stealth" bomber of the Eighties.

In combat, however, only the performance figures were equal. Despite the fact that Sabres were usually outnumbered three or four to one, the victory ratio came out in their favor: 13 MiGs killed for every Sabre lost. The reasons are fascinating. First, as always, was pilot quality and aggressiveness. The USAF sent in its postgraduate team, while the North Koreans—actually the Warsaw Pact nations and the Chinese, for the most part—used such combat to season their pilots, perhaps even to winnow out the less adept for the future good of the service. Second, the F-86 was a superior gun platform, able to maintain its attitude and bank angle at altitude. The ship had its special all-flying tail to thank for this. The "stabilator" (combining the functions of horizontal stabilizers and elevators) rode part way up the vertical stabilizer, rather than low and in-line with the wings.

Naval air power developed in a somewhat more deliberate fashion. Special requirements for structural strength and lower approach speeds for carrier landings influenced the pace of development. Nevertheless, fighters of interesting mien cropped up. These included the Phantom and Banshee series which established McDonnell as a dominant manufacturer for the future. Grumman Cougars appeared, along with such exotica as the tailless Vought Cutlass.

Myriad developments emerged, including specialized fighters such as the McDonnell XF-85 Goblin, designed to be launched from a trapeze beneath a B-36 bomber, and the big bruising Curtiss XP-87 Blackhawk. This last item, a four-engine fighter, was originally proposed as an all-weather attack airplane. Bad luck and indifferent performance tainted its career. Though a superhot airplane, the lovely Martin XB-51 was burdened with so much political baggage that it never reached production. In 1951, however, the Martin B-57 Canberra, a modified English Canberra, did go into service with the USAF. Another airplane was brought into service for one purpose, and then, 25 years later, found active employment in a second career. The Douglas B-66 and its Navy A3D counterpart served as bombers for a while, but found longer life as tankers and electronic countermeasures aircraft.

This first jet generation was ambitious. The first attempt at a supersonic fighter, the Republic XF-91, flew on May 9, 1949. Design began on the North American XF-100, the first operational supersonic fighter, in January 1951. It went supersonic during its first flight on May 25, 1953. Development lead times were shorter in those days, but no less troublesome than today; it would be April 1955 before the first F-100 squadron became operational, flying F-100Cs. Accidents had led to numerous structural and aerodynamic changes in the original models.

Four of the fighters that would figure in the following decades and serve in the Vietnam War made their first flights during this first wild period of development. Each represented a novel approach to different military problems, once again illustrating the richness of both the American military budget and the engineering resources that have backed it up. All served with great success, though not always in the role originally intended for them.

The first was the McDonnell F-101 penetration fighter, designed just after World War II, which first flew on September 29, 1954; by then it was a long-range interceptor and a reconnaissance plane. The delta-wing Convair F-102 came next. Equipped with a Hughes fire-control system and sophisticated missiles, it and its follow-on, the F-106, would

predominate in American-based interceptor squadrons into the Eighties. Clarence "Kelly" Johnson's tiny, blindingly fast Lockheed F-104 was not deployed widely in the USAF but met European needs brilliantly, and was the standard fighter of a number of foreign air forces for many years. Finally, the "Thud"—the Republic F-105—went through a long and painful development process (like the F-101) before emerging as a workhorse in Vietnam.

Bombers were developed in less profusion, but their capabilities exceeded all ambitions and expectations, especially the Boeing B-52. When the YB-52 first flew on April 15, 1952, nobody could have imagined that the production model would remain the frontline strategic bomber into the Eighties. Indeed, it may still play a significant role into the next century.

Similar advances were made by the air forces of other nations during this extraordinary decade, which for reasons of engineering and economy will probably never be matched in a similar time period.

Though the U.S. Air Force met technical competition from many countries, the United States reigned supreme in civil air transport in the years after World War II. The prewar success of the DC-3 and wartime development of the much larger, faster Douglas C-54 (DC-4) and Lockheed C-69 (Constellation) kept the United States in front of all other nations, and the expanding air transportation economy reinforced the U.S. lead. Boeing's rotund C-97 transport found itself transmogrified into the pot-bellied, though dignified, 377 Stratocruiser. And while passengers loved its speed and luxury, it was not as efficient as the next generation of piston-engine airliners, so relatively few were made. Yet, these aircraft

WWII engineers designed the Convair B-36, top, as an ultra-long-range bomber of huge proportions—a precaution in case Britain fell and the Allies needed to bomb Germany from North America. Eventually fitted with four jets in addition to its six piston engines, this largest bomber of all time served in the U.S. Strategic Air Command until eclipsed by the sweptwing B-47 jet. An immense tire dwarfs a B-36 mechanic, above.

B-47 "The One to Fly"

Walter J. Boyne

When it appeared in 1947, the B-47 represented a tremendous advance in the state of bomber art, combining 35-degree sweptwings with six jet engines in four underwing pods. The engines were among the first generation of jets, but were reliable, though extremely sensitive to starting procedures and throttle manipulation.

Perhaps the best part about flying the B-47 was being a member of the Strategic Air Command's first team. The term hadn't been coined yet, but there was a tremendous Cold War bomber gap—all in our favor—and that made the B-47 *the* airplane to fly.

Flying the B-47 was wonderful. It was a responsive aircraft, very maneuverable, and receptive both to control inputs and throttle corrections.

Besides its load of crew, cargo and bombs, it also carried a wild baggage of horror stories about its dangerous characteristics. You could get in trouble in a B-47 if you weren't careful, but you can in any airplane, and few are as rewarding to fly as the B-47.

Takeoffs at a high gross weight, a field elevation of more than 4,000 feet and a high outside temperature could be tricky. It was a daunting experience to sit with the throttles full up, feeling little acceleration as the runway markers slid slowly past, watching the end of the runway come up. Losing an outboard engine late in the roll was really hazardous. But it almost always made it, and once airborne, the gear could be raised, the flaps milked up and an exhilarating 310-knot-climb established.

The visibility from the clear canopy was superb, and while the missions were often grueling—some as long as 24 hours—they were so jammed with activity that time passed quickly. The B-47 was immensely capable, with an excellent bombing system and speeds that made interception by the fighters of the time improbable.

Landing the B-47 was both demanding and enormously satisfying; it could be brought down whisper soft, the front and back trucks of the gear touching simultaneously.

The B-47 (it was always called just that, never the "Stratojet" or anything else) required skill and training of a high order, but that was part of its wonderful mystique.

A B-47 Stratojet medium-range bomber blasts off from Edwards Air Force Base with a boost from JATO (Jet—actually rocket—Assisted Take Off). Boeing risked millions to develop the first service bomber with narrow, swept-back wings and underslung pods for jet engines.

Air-to-air refueling of the B-47 Bomber, here by a Boeing KC-97, enabled the USAF Strategic Air Command (SAC) to fulfill its mandate—to maintain peace through deterrence. First introduced to the SAC in 1951, the sophisticated bomber could span oceans and continents at near supersonic speed.

dominated the world's airways during the late Forties, making nonstop transcontinental flights and eventually transatlantic flights enormously more popular.

Makeshift modifications transformed a British four-engine bomber into the Avro Lancastrian, to join the York and Tudor aircraft; only the last was even moderately successful. France quickly responded with the SNCASE Languedoc and the SNCASO Bretagne. Like the more ambitious aircraft that followed, these served to bolster national pride, but were not efficient enough to generate the kind of profits the American aircraft did. It was not a question of ambition or a lack of vision; some of the largest aircraft of the period came from Britain, including the eight-engine Bristol Brabazon and ten-engine Saunders-Roe (Saro) Princess flying boat. France responded to the quest for size with the SNCASE Armagnac and Breguet Provence. But these makers could not integrate the factors involved in truly efficient design, manufacture, marketing and operation of passenger aircraft. The British and French aircraft could not be produced in sufficient quantity to drive the unit cost down or to achieve the low seat-mile costs that generated profits. Nor could their companies provide the spares, maintenance and service of the polished American firms.

American aircraft kept improving. The airliner manufacturing business came to resemble a high-stakes poker game in which one company would make improvements, betting on an expanding market. Other companies would see the bet and raise it. The Lockheed Model 049 Constellation (C-69 in military service) was stretched, streamlined, reengined, given additional fuel capacity for greater speed, range and passenger capacity and evolved through dozens of military and civil designations. In direct competition, the Douglas C-54 evolved into a civilian series, including the DC-4, DC-6, DC-6A, DC-6B, DC-7, DC-7B and DC-7C. The C-118 series became military craft.

For all aircraft manufacturers, the Holy Grail was the fabled "DC-3 Replacement." Everyone thought that postwar technology would be able to produce an aircraft faster, better, more profitable and more durable than the 1935 DC-3. Companies brought out one design after another. There was a mid-wing arrangement with twin engines, a low-wing craft with four small engines, high-wing and trimotor combination, a mid-wing with buried engine and propellers at the rear and, finally, four engines coupled and buried in the wing to drive two propellers—nearly every conceivable configuration. Everybody failed, even Douglas with its Super DC-3.

Twin-engined aircraft did have a role in transportation, however. Cost factors were not so vital as in the four-engine Constellations and DC-6 series. National pride dictated the employment of indigenous airliners on home routes. America entered the field with the Convair CV-240 and Martin 2-0-2/4-0-4 series. Sweden's long-lived Scandia was between the Convair and the DC-3 in size, and the Soviet Union countered with its series of twin-engined Ilyushin transports. Britain responded with the beautiful Airspeed Ambassador and the dowdy, doughty Vickers-Armstrong Viking.

But Vickers learned fast; its Viscount put Great Britain back into the role it perceived as proper—supplier of aircraft to the world. Quiet, comfortable and fast, the four-turboprop (a gas turbine akin to a jet engine but driving a propeller) Viscount began service on July 29, 1950, on that famous proving route, London to Paris. European and American airlines vied for a position on the Viscount delivery line. In those days, one of life's special pleasures was to fly in the rear of a quiet, all but vibration-free Continental Air Lines Viscount and be plied with champagne by the stewardess. Also in Great Britain, Handley Page produced the Herald and Avro, the 748, both turboprops that would soldier on for many years. Fokker took the lead with its long-lived, handsome, economical F27, which was produced both in the Netherlands and in the United States by Fairchild.

The Soviets drew from design features of their bombers to create the Tupelov Tu-104 and later the Tu-124V. After the British Comet 1s, the world's first all-jet airliners, were introduced in 1952 and then withdrawn from service, the Tu-104 was the world's only operational jet passenger craft for a time. Though not sophisticated by Western standards, the Tu-104s served the U.S.S.R. well as interim aircraft pending the development of later, more specialized types. Serving immense territory of the Soviet Union as well as overseas routes, the Soviet State-owned monopoly, Aeroflot, soon became the largest single airline in the world.

Meanwhile, many other more revolutionary aircraft flourished in the postwar era. The long-held desire for vertical flight was realized. The intrigue began with Leonardo da Vinci and his "airscrew" and continues to this day. The balloon had drawbacks that a heavier-than-air craft could remedy. The obvious convenience of vertical takeoff and landings was more than offset, however, by the complex technical problems designers faced. Extremely short flights were made by experimenters in various countries, but none of their designs were practical.

For a time, Spanish inventor Juan de la Cierva's introduction of the autogyro, a wingless aircraft with a helicopterlike rotor, bypassed the problem almost entirely. As an autogyro moves through the air, its unpowered rotor spins—or autorotates—in the wind stream, thus performing a wing's lifting function, while a single engine and propeller up front provide propulsion. Orville and Wilbur Wright had been the first to realize that a propeller—or rotor, in the horizontal plane—was, in actuality, a rotating wing. Applying this principle, autogyros achieved comparatively short takeoff and landing distances; the greater the wind, the shorter the distance. However, because the rotor was not powered, such craft could not climb or descend vertically, or fly backwards or sideways as can helicopters.

The Germans made perhaps the first public demonstrations of a fully controllable helicopter in 1936, with the Focke-Achgelis Fa 61, a twin-rotor that established many records. Hanna Reisch, a superb test pilot, first revealed the craft to the public by flying it into a roofed sports arena in Berlin in 1938. The contribution of the Focke-Achgelis was limited; it did not give rise to succeeding generations of better helicopters. Surprisingly, the war did little to spur helicopter development in Germany, and although the world's first military helicopter, the German Flettner Fl 282, saw limited deployment as an antisubmarine observation aircraft in 1942, few were manufactured.

In the United States, helicopter development was practically the exclusive province of Russian aircraft designer Igor Sikorsky. His first successful machine, the VS-300, flew in the summer of 1939, beginning a dynasty of helicopters that continues to this day. Competition from Bell and Piasecki and others added zest to the ensuing commercial battles.

Development soared after World War II. Ever more powerful engines and improved transmissions supported a frenzy of experimentation. Not many companies entered the market, though dozens of "one-off" types by enthusiastic inventors caught the public's eye. A jet helicopter—the von Doblhoff

Intended as a flying oceanliner, the elegant British Saunders-Roe "Saro" Princess flying boat, left, was already an anachronism by the time of its release in 1952. The success of more efficient long-range land-based aircraft made it obsolete before it flew. Howard Hughes' HK-1, bottom left and right, shared the same fate as the Princess. Popularly termed the Spruce Goose, *the U.S. craft had, at 320 feet, the longest wingspan of any airplane ever built. It was intended to supply wartime Great Britain by air. Finished too late to fulfill its mission, the eight-engine oddity took only one short hop in Los Angeles harbor in 1947 and then headed for mothballs. The* Goose *was mostly made of birch plywood.*

Evolution of Fighter Tactics: Korea

The Korean Conflict of the early Fifties was the first jet air war. Surprisingly, this innovation initially changed air-combat tactics very little. The weapons were much the same, machine guns and automatic cannon, but the combat ceilings continued the trend begun in World War I by climbing to well above 40,000 feet.

Although Fighting Wing tactics and Finger Four formations (see page 152) were still the vogue, the distance between fighters was increased (to about 500–1,000 feet between wingmen, and up to 8,000 feet between elements) in compensation for the much faster speed and larger turn radius of the jet fighters.

The short respite between wars had also allowed for some refinement of coordinated fighter tactics, resulting in closer teamwork between the two sections of each division. Now each section typically alternated attacks on an enemy formation in the pattern often called Fluid Four. The Communist air forces in the Korean conflict employed very similar tactics, but usually combined more two-plane Fighting Wing sections to create larger formations.

ROBERT L. SHAW

The High/Low Split-*(Korean War) "MiGs Alley," North Korea, 1952: Two American F-86 Sabres (upper right) engage in high speed pursuit of two Chinese pilots in North Korean MiG-15s (center). Both MiGs coordinate evasive action, one "zoom" climbing (upper left), the other breaking left and diving (lower right). Unable to match the MiG's rapid rate of climb, both Sabres pursue the diving MiG. The climbing MiG rolls toward the Sabres, diving with room to spare (center), and scores hits with its 23mm cannon. The stricken American's canopy pops off as he prepares to "punch out," or eject.*

WNF 342—had been flown by the Germans in 1942, but not until modern, lightweight jet engines arrived did the helicopter gain sufficient—and relatively vibration-free—power to cause a boom in vertical flight.

The saga of aircraft design is often glorious, tragic, triumphant or bitter, but rarely has it been so poignant as in the case of the first pure-jet transport. Airliners with turboprop engines had provided Great Britain with the means to ease back into the international aircraft market. By parlaying the success of the Viscount and by quickly commercializing pure turbine transport, the British believed that they might wrench leadership in airliner sales away from the American manufacturers.

Aesthetically, few aircraft have been more attractive than the de Havilland D.H. 106 Comet 1. Years ahead of its time in appearance and performance, the aircraft was also years ahead of the aeronautical engineer's knowledge of metal fatigue and its potentially devastating effect on the pressurized cabins of jet aircraft. From the D.H. 2 of World War I to the Mosquito of World War II, de Havilland had produced many successful, innovative aircraft—sometimes unorthodox but always in the forefront of aviation. Perhaps only a firm with such a background would have attempted to jump ahead of all the world's manufacturers to create the first jet airliner, and take away America's remarkable hold on the free world airliner market.

The magnificent Comet 1A entered commercial service on the London to Johannesburg run on May 2, 1952. Carrying 36 passengers at 500 miles per hour—and with trouble-free regularity—the Comet 1A was a worldbeater, and airlines began to queue up to buy.

Disaster came when two Comets broke up in flight, the first on January 10, 1954, the second on April 8, 1954. Engineers isolated metal fatigue as the problem; the aircraft's fuselage had been inadequately designed for repeated pressurization, especially since the Comet cruised at higher altitudes than other airliners of the day. De Havilland tried to recuperate with a reengineered version, but the very name Comet had become an anathema to the traveling public.

Like de Havilland, the Boeing Company had a long history of successful gambles that paid off in brilliant aircraft. But unlike de Havilland, it had considerable experience in building large jet aircraft; with its B-47 and B-52 bombers providing endless hours of engineering data. Still, considerable financial risk was involved in the creation of the world's first successful jet transport, the 707, and the similar Air Force KC-135 tankers.

The 367-80 prototype of 1954 (immortalized at Boeing as the Dash Eighty) was as much an improvement over the ill-

A trio of Russian-built MiG-15s cruise "MiG Alley," a corridor of North Korean territory that saw most of the air-to-air combat of the Korean War. MiG-15s found a formidable opponent in the American F-86 Sabre, opposite, largely because of superior U.S. pilot experience and training rather than great disparities in aircraft performance. By war's end, F-86s had recorded 827 confirmed victories over MiG-15s with the loss of only 78 Sabres, a kill rate of better than 10-to-1.

fated Comet as the Comet had been over piston-engine transports. In fact, the resulting 707 airliner came as close as economically possible to Mach 1, the speed of sound. Even today none of its subsonic successors are notably faster and, because of the constraints imposed by the need for fuel economy, many routes are flown at the same speeds initiated by the 707.

From the financial point of view, the transition from piston-engine power to pure jet was not without its problems. Airlines were not immediately ready to pay for the delivery of the sleek but large and expensive jets. An Air Force contract for a number of KC-135s (the military tanker equivalent of the 707) permitted Boeing to weather this storm, however, and the company soon received more orders for the 707 than it could fill.

A series of triumphs followed, beginning with the three-jet Boeing 727, still the most-produced jet transport in history. Later, the Boeing Company introduced the first and most successful wide body, the 747, as well as the series of 737, 757 and 767 aircraft, all noted for their fuel economy. Other firms learned from Boeing's book. Douglas produced the 707 look-alike (but quintessentially Douglas) DC-8, while Convair built the ambitious but commercially unsuccessful 880 and 990. All around the world manufacturers tried to emulate Boeing's example.

The 367-80 (a $16,000,000 project for Boeing) has probably generated more revenue than any other prototype in history; its descendants have provided perhaps the most reliable offset to the U.S. balance of payments problems. But its contribution has been far more profound than mere aerodynamics and economics. The 707 and its brethren revolutionized transportation: they gave mankind mobility hitherto undreamt-of. Shuttle diplomacy began with the 707 type, and economic vacations to distant lands became possible for people of every culture.

Great Britain employed the completely redesigned and enlarged Comet 4 on its airlines, and a few others purchased the new ships. The French were not far behind the original Comet, flying their smaller, twin-engine Sud-Aviation Caravelle on May 27, 1955, only two years after development started. The Caravelle used both the nose section and the engines from the Comet to ease development efforts, and the airplane was an instant success. Part of its appeal lay in the then unusual mounting of the two jet engines aft on the fuselage. The sleek, exotic ship had great passenger appeal over short routes, where it was most economical. Caravelle cracked the American market—a totally unheard of event—and 280 were sold around the world.

Optimism helped to fuel the major new market in large, turbine-powered airliners. All over the world, companies invested large sums of money in the development of aircraft, some of which ultimately did recover their costs.

Among the British de Havilland company's unique designs of the early Fifties: above, a Vampire twin-tail jet fighter and trainer sit before the Comet, the world's first commercial jet transport. After several Comet crashes, de Havilland hurried to repair structural flaws, opposite, but the public shunned the aircraft. At left, Princess Elizabeth descends from a Comet. She has returned from a Commonwealth visit at word of the death of her father, King George VI.

COMET SERIES II
SERIAL No 06045.
PROJECT No 5566.

Such a potentially risky approach has always characterized aviation. In earlier times, though, a few thousand dollars and a few weeks were enough to create a prototype. In the new and sporty game of high finance, firms that developed big four-turboprop cargo planes or four-jet passenger liners literally invested years of preparation and hundreds of millions of dollars in any new offering to the market. The manufacturer bet his company's total resources and future on the new craft.

It took some time for the market to grow to meet the opportunities offered by the big new jets, which carried at least twice as many passengers as their propeller-driven predecessors at almost twice the speed. Hot-blooded advocates of the new types often saw markets where they did not exist. Perhaps the most elusive target was the cargo jet; the need was always proclaimed, but in the last analysis, the available cargo space on the existing fleet of passenger jets was more than ample to handle most of the demand. Not until the advent of such farsighted operations as Federal Express did overnight delivery of cargo become a lucrative industry. They devoted and dedicated their entire operation to the shipment of priority items.

Visions of aerial freighting, combat, intercontinental travel or even of cross-country flight were not the only factors influencing the development of aircraft. A more modest airplane—the lightplane—was sought, a machine that would enable people to putter around in the air. Such birds emerged in various places around the world after World War II, but had really come into their own in the United States during the late Thirties.

Lightplanes were typically minimal aircraft, with small engines, simple airframes and speeds that hovered around 75 miles per hour. But even these small machines cost a good bit more than automobiles and were infinitely more complex to operate. Americans equated the Piper Cub with the Ford

The Douglas DC-7C, above, one of the aircraft industry's highest developments in piston-powered airliners, was built to cross the Atlantic in either direction. It began service with Pan Am in 1956. By the mid-Sixties, the Boeing 707 and other jets were replacing piston-engine liners. Boeing's $16,000,000 gamble, right, the 367-80, was the prototype for the remarkable 707. Experience gained on the B-47 and B-52 bombers paved the way for the new civilian craft, and the Air Force bought the first batch as KC-135 Stratotankers in 1954. By 1967, 568 Boeing 707s operated commercially with such airlines as Pan Am, Quantas and American.

Model T. The press consistently told the public that it should be able to fly as easily and conveniently as it drove, and not for a great deal more money.

While this dream ignored the realities of flight, the belief that someday there would be an airplane in every garage never dimmed. This expectation extended even to government levels. In November 1933, Assistant Secretary for Air Commerce Eugene Vidal held an elaborate contest to create a $700 Everyman's Airplane. He did this against industry protest. It was an impossible dream; the eventual winner, the twin-boom Stearman-Hammond, had a Menasco engine that alone cost $7,000!

World War II had a profound effect on the lightplane industry, or general aviation, the term for business and pleasure aviation. Pilots were trained on a mass basis, for instance, beginning with the Civilian Pilot Training Program and extending into the enormous buildup for the armed services. At war's end, ten times as many pilots as ever before entered the civilian economy. Many manufacturers believed—or hoped—that they were all going to insist on owning their own airplanes.

The conversion from wartime to peacetime footing in the lightplane industry was chaotic. Piper, Stinson, Taylorcraft, Aeronca and others geared up for the prophesied boom, and dozens of smaller firms leaped onto the gravy train.

For one splendid year of monumental production the good times rolled. More than 17,000 lightplanes had been produced by 1947. For the first time in history a boom market ruled the private aircraft business. Then sales fell dramatically. Many wartime pilots apparently wanted to get on with college and raising families and getting started in careers. Others had had enough of flying to last a lifetime.

Even at the height of the boom, industry standards of aircraft manufacture remained high. Few would have be-

The Area Rule for Supersonic Fighters

Richard T. Whitcomb

Aerodynamicist Richard Whitcomb demonstrates his revolutionary design breakthrough in the wind tunnel at NASA's Langley Memorial Aeronautical Laboratory in the early Fifties. When Convair redesigned the fuselage of the F-102, left, to resemble the shape of a pinched-waist Coca-Cola bottle, opposite, the aircraft could easily break the speed of sound—an astounding speed gain of 20 percent over the old, straight-waist design.

During the late Forties, the development of axial flow jet engines with high thrust led to a concerted, national effort to achieve supersonic flight for military aircraft. Wind tunnel and rocket-powered model tests, though, indicated that drag in the form of strong shock waves at supersonic speeds for many of the proposed supersonic airplane configurations was greater than the maximum thrust of their engines. Systematic research conducted in the new transonic wind tunnel at the Langley Aeronautical Laboratory of the National Advisory Committee for Aeronautics (now part of NASA) suggested that much of the increase in drag came from adverse interactions of the flow fields produced by such airplane parts as the fuselage, wing and tail surfaces.

To learn about these interactions, I conducted studies of the flow fields around a simple but representative sweptwing-fuselage combination in the transonic wind tunnel. A scale model was used. The shock waves produced by the configuration were observed using a Schlieren optical system that makes such waves visible. The results were surprising and intriguing: instead of the expected—two separate shock waves produced by the wing and fuselage—only one strong shock wave existed, just aft of the wing. It was very similar to that produced by a fuselage alone. It seemed that the mathematical conversion of the complex configuration to a single, simple fuselage shape might help to predict a shock wave's drag. But what longitudinal shape should this fuselage have?

In December 1951, while pondering these results, it suddenly occurred to me (much like the proverbial light bulb over a person's head in a comic strip) that the shock wave and the associated drag for the fuselage-wing combination is the same as that for a simple fuselage alone that has the same longitudinal variation of cross-sectional areas as that of the fuselage-wing.

The concept immediately explained the excessive transonic and supersonic drag for many military aircraft designs. More importantly, from a practical viewpoint, it suggested a method for eliminating this adverse effect and even obtaining a favorable effect. This could be accomplished by shaping the fuselage in the vicinity of the wing to reduce the total cross-sectional area. Such changes result in a fuselage with a pinched waist, like an old Coke bottle.

An extensive, systematic investigation was then conducted in the transonic wind tunnel that provided overwhelming, definitive proof of the concept. The results of this investigation were transmitted immediately to the aircraft companies designing supersonic military aircraft. On the basis of this concept, substantial reductions of transonic and supersonic drag were achieved.

USAF

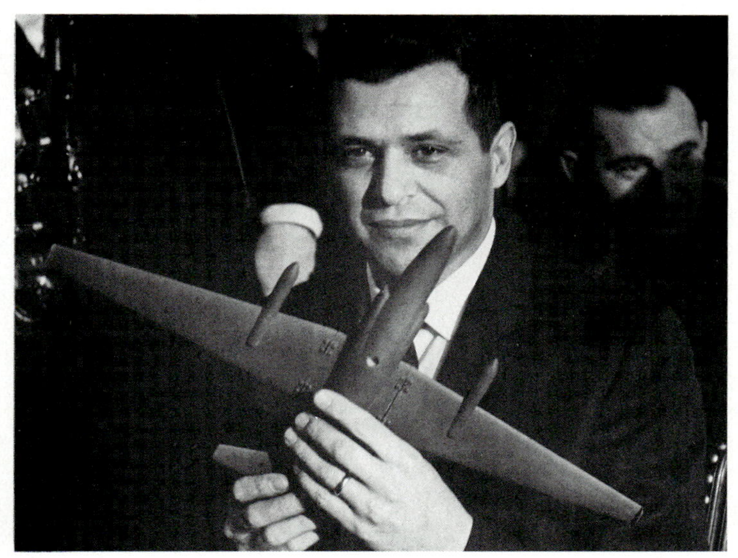

Dreadnought of the skies, the B-52 Stratofortress, the USAF Strategic Air Command's bomber is affectionately called BUFF (Big Ugly Fat Fellow) by its crews. The world's largest strategic bomber, it can deliver armaments more than 8,000 miles without refueling. A U-2, shown as a model with its most famous pilot, Gary Powers, above, was shot down over Soviet territory on May 1, 1960. The incident, with its diplomatic backlash, highlighted the value placed on overhead reconnaissance in the Atomic Age.

lieved that many of those 17,000 aircraft would still be around 40 years later, yet many aircraft do survive from that vintage year of 1946. Still in good condition and available for sale at affordable prices, they compete in the marketplace with used airplanes from every year thereafter. Remarkably similar to current models in style, safety and performance, this backlog of used aircraft—and legal liability problems—has effectively ended new lightplane production.

Just as the Boeing 707 plumbed the upper limits of jet transport performance right from the start, so did the postwar crop of general aviation aircraft set performance standards that would not be substantially improved during the next ten years—or the next forty. The Beech Bonanza appeared with its distinctive V-tail styling and immediately displaced the prewar Staggerwing Beech as the most desirable general aviation aircraft. Today, the Bonanza remains, despite the introduction of other favorites such as the North American Navion and Globe Swift, the clear leader.

Performance and styling have remained fairly static since the Forties. No radical advances appeared; Cessnas look like Cessnas and Pipers look like Pipers—and the buyers looked away. Finally, in the Seventies, the twin-engine business aircraft, equipped with better radios and instruments, provided genuine utility for an emerging market and began to pull general aviation out of its self-imposed doldrums.

By 1955, conditions were ripe for a new spurt in aviation growth in every field. This next growth—and every succeeding period of expansion—would be more rigorous in every regard, from the capital required to the level of engineering to the heights of pilot skill and daring.

Bigger, Better, Faster

Relatively few of today's expensive aircraft reach buyers, whether commercial or military; those that do are highly sophisticated. In the sense we use here, sophistication is almost synonymous with specialization: but however these two words are defined, they add up to astronomical price tags.

Aircraft and their many accessories have always been relatively expensive, as have the services of the people who build and maintain them. Since the Sixties, though, costs have skyrocketed while aircraft production has plunged. Boeing's B-17 of World War II cost $187,742 per copy, reflecting great economies of scale as nearly 13,000 copies rolled off the production lines. By the mid-Eighties, the U.S. could afford to build only 100 B-1B bombers at a cost of hundreds of millions of dollars each. At one time, Piper Cubs sold for as little as $1,325; by 1986, so great had been the increase in various costs that no major manufacturer built single-engine light aircraft.

A number of factors affected aircraft design during the Seventies and Eighties, but perhaps the most important was the great rise in fuel prices following the Arab oil embargo of 1973. This crisis of the global marketplace caught the West-

SR-71, left, can maintain speeds of Mach 3 and fly at more than 80,000 feet. Successor to the U-2 (and also designed by Lockheed's Clarence "Kelly" Johnson at the Burbank, California, "Skunk Works"), the billion-dollar "Blackbird" features a basic structure of titanium and high-temperature plastics. The AV-8A Harrier fighter-bomber, below, flown by the U.S. Marine Corps, can take off and land vertically, hover and exceed 550 miles per hour in the straightaway. Overleaf: BAC-Aérospatiale Concorde receives preflight service. Its hinged nose droops, ensuring a clear pilot's view of the runway during ground operations and takeoff. For supersonic flight, the nose swings upward.

Speed and Payload

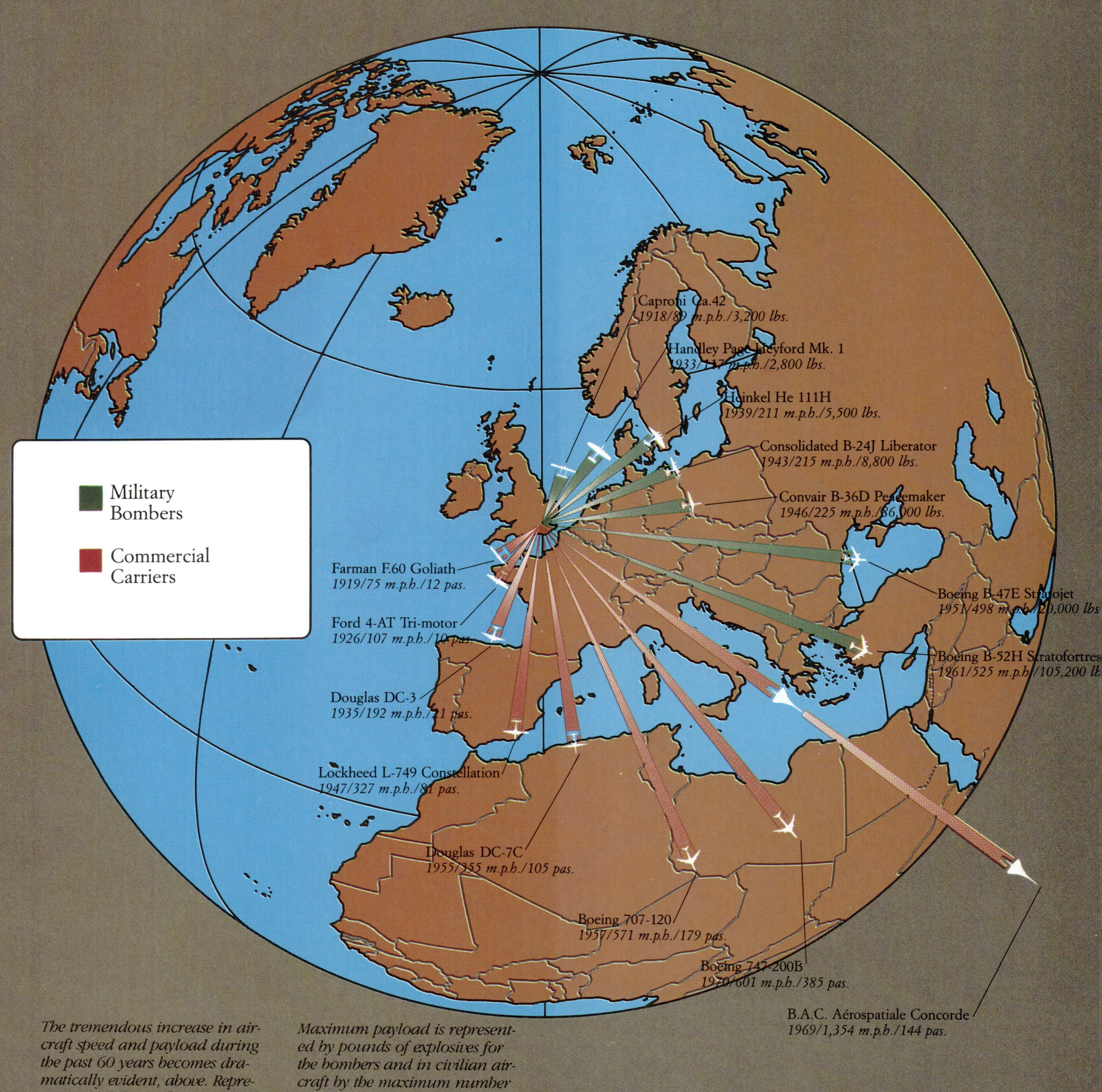

The tremendous increase in aircraft speed and payload during the past 60 years becomes dramatically evident, above. Representative military (bomber) and commercial (passenger) aircraft from each era illustrate the average distance covered in three hours flying time at cruise speed.

Maximum payload is represented by pounds of explosives for the bombers and in civilian aircraft by the maximum number of passengers. The Concorde is an exception, shown at only two hours of flight time—its speed and fuel load have limited its range.

Outracing its contrails, the BAC-Aérospatiale Concorde streaks across the sky during an annular eclipse of the sun.

Painters at Lockheed spray an L-1011 TriStar, a widebodied jet suited for cross-country service, below. Right, inspectors at Sikorsky Aircraft carefully check a composite material for aircraft that includes Kevlar and graphite fibers bound by epoxy adhesive. The aircraft industry maintains strict standards of quality control, especially with parts and materials subject to stress.

ern world by surprise. And yet while aircraft prices have soared, the principal user of an aircraft may actually find that costs have *fallen*.

Aviation's price-cost paradox reflects a complex web of economic and political factors. Deregulation of the airlines has greatly increased competition among the regular carriers. Flying used equipment for the most part, new companies have emerged with pricing practices that drive older, more famous airlines out of business. While passengers enjoyed artificially low ticket prices, they also discovered the anxiety attending sometimes questionable maintenance practices.

In the military, higher costs are usually the result of design changes that seem to add to the versatility of an airplane. The giant Lockheed C-5A transport provides an example. Serving faithfully since its introduction in 1968, it has performed very well within the limits of its design, but the press has treated it as a failure because of cost increases and a shortened life span for the wing structure.

The C-5A's difficulties arose because Air Force program advocates adorned it with artificial requirements; Lockheed had to build these into the design, making numerous technical compromises necessary. The original concept of the C-5A called for a very heavy transport that could carry any item in the Army's inventory, including tanks.

The C-5A became a victim of a hobby-shop mentality, in which C-5A advocates saddled the design with requirements that would rarely if ever be needed. A huge aircraft, the C-5A's wingspan was 222 feet, its empty weight 321,000 pounds, and maximum weight at takeoff of 769,000 pounds, nearly 385 tons. Yet the advocates insisted that a heavy and immensely complex landing gear be designed into the airplane so that it could be used from semiprepared fields. The odds were astronomical against so large an aircraft ever being deployed from anything but standard, paved runways, yet the designers had to meet the challenge.

An elegantly complex crosswind landing gear was designed so that the tire pressure could be changed automatically in flight to suit the landing conditions; the landing gear also could be made to "kneel" so that cargo could be more easily transferred from truck to the airplane's cavernous cargo bay. As a landing gear it was a landmark of theoretical universal utility. However the multimillion-dollar C-5A was far too precious ever to risk on a soft field. It has landed throughout its career on the same long, wide and smooth runways that passenger airliners use. It all looked great on paper, but this added landing capability cost dearly in terms of weight, range and maintenance.

The engineers had been outbid by the hobby shoppers. The addition of weight, the greatest enemy of any airplane design, was so significant in the C-5A that designers had to skimp on the main structures of the airplane to reduce the weight. The final product had wonderful performance and

A heavily suited Pratt & Whitney technician runs part of a jet engine through a carefully controlled heating and cooling process. The new technique gives super strength to metals, helping jet engines withstand high internal temperatures and stresses.

Evolution of Fighter Tactics: Vietnam

The late 1950s and 1960s brought a series of local hot spots in the "Cold War," among them conflicts in Vietnam and the Middle East. This period also saw the advent of the next two technological innovations to have a significant impact on air-combat tactics, the air-to-air (AAM) and surface-to-air (SAM) missiles.

The effect of the quantum increase in effective range of the AAM over the gun was to render the Fighting Wing formation (see page 152) obsolete, since close aircraft spacing limited the range of visual coverage of the section's rear.

ROBERT L. SHAW

The Barrel Roll Attack—*(Vietnam War) west of Nam Dinh, North Vietnam, 1972: An American Navy F-4 Phantom II (center) engages a North Vietnamese MiG-21 (upper right) at long range as it crosses his flight path. The MiG turns left, oblivious to the pursuing F-4. Rather than overshoot him the Phantom climbs, executing a "rolling pull" away from the MiG. Inverted to watch the MiG at the top of his climb, the American maintains his distance behind the North Vietnamese pilot "in the vertical." Rolling into his dive, he completes a 270-degree turn and the Phantom is in perfect position to fire a Sidewinder heat-seeking missile that will fly into the MiG's tailpipe.*

versatility, but was severely compromised in structural strength and longevity.

Despite all the complaints about the increasing complexity and expense of military aircraft, at mid-century new types flowered all over the world and blazed new trails in every category. Delta-winged aircraft, including the Convair F-102 and B-58, vied with such straight-winged "missiles with men in them" as the Lockheed F-104 Starfighter. During the constant quest for high speed, reasonable takeoff and landing performance, "variable geometry" aircraft appeared with wings that could be swung out during landing and swept back for high-speed flight. Great Britain produced hot prototypes like the TSR.2 while France concentrated on her best-selling line of Mirage fighters. The U.S.S.R. continued to produce clouds of aircraft—a real aluminum overcast—including the world's most advanced line of bombers, all bearing Western code names like "Blinder" and "Blackjack." Jet engines throughout the world became more efficient and powerful, far surpassing their piston-driven ancestors in only a short time. Reliability rose, at first slowly, then dramatically, while specific fuel consumption declined.

Aircraft of the Fifties and Sixties all went farther, higher and faster than their predecessors; fire control, bombing and navigation technology matched progress in airframe components and engine performance. The computer came to the aid of the designers and pilots of these hot ships. Sometimes, in a reverse instance of the C-5A case, money poured into research and development yielded unexpected returns or benefitted later generations of aircraft. Consider the two ultra-advanced North American products, the Navy's A-5 Vigilante and the Air Force XB-70. The Vigilante was a victim of changing strategic nuclear policy but turned out to be an outstanding reconnaissance type. The high-flying XB-70 had its mission jerked out from under it by the same high-reaching Soviet missiles that downed Gary Powers' U-2. Each of these aircraft provided practical flight testing of aerodynamic design, new materials, electronic systems and manufacturing techniques that are still in use. The bill for many current aircraft's development has been significantly reduced by the research and development surrounding these two pioneers.

The search for performance sometimes gave unexpected results. For instance, after fighter designers had brought forth aircraft capable of Mach 1 (the speed of sound, a speed that varies with altitude but which approaches 760 miles an

A Soviet built MiG-21 with droppable fuel tanks and AA-2 "Atoll" air-to-air missiles engages a USAF Republic F-105 Thunderchief over North Vietnam. Red victory stars on the MiG's nose reveal an experienced pilot, perhaps one of North Vietnam's premier aces such as Captain Nguyen Van Bay or Colonel Tomb.

Combat artist William S. Phillips painted "Heading for Trouble," above, featuring two AH-1G gunship helicopters from F Troop, 4th Cavalry following their scout, an OH-6A "Loach," in a pre-dawn search-and-destroy mission. A CH-46 Sea Knight, left, and other helicopters in Vietnam rushed wounded soldiers to medical units. In future conflicts, combat and rescue missions may be further shortened by the deployment of advanced types. A Sikorsky X-wing experimental craft, opposite, takes off with a rotor that becomes a fixed wing for fast flight. Jet engines provide thrust.

Cutting in his afterburner, an F-4 Phantom II pilot suddenly doubles his acceleration. Added boost or "wet thrust" comes from injection of fuel into hot exhaust gases. Much of the Navy Phantom's heavy action during the Vietnam War was coordinated by an earlier version of today's Grumman E-2C Hawkeye, left. Like a radar antenna 30,000 feet high, the carrier-based Hawkeye can "see" nearly 300 miles, greatly extending the "eyes" of the carrier. A ground control intercept officer operates a sophisticated radar system during Red Flag—war games held near Nellis Air Force Base in Nevada.

hour at sea level) they started to consider the next generation. If Mach 1 was nifty, Mach 2 should be a double delight. So, fighters capable of Mach 2 and more appeared all over the world. Vietnam, however, taught that aerial combat usually took place at around .9 Mach or less. Fuel consumption rose dramatically at higher speeds, as did pilot fatigue. Furthermore, air-to-air missiles could always be made to outrun airplanes, giving maneuverability precedence over speed. Mach 2 simply wasn't necessary for combat.

Boosts to performance brought about an increase of attendant ground equipment, though. The maintenance time required per hour of flight expanded tenfold as sophisticated radars, gun and missile guidance equipment and communication and navigation devices were incorporated into already cramped airframes. Each new system required technical manuals, test equipment, calibration equipment, spare parts channels and specially trained mechanics. These systems (often called black boxes) frequently required more maintenance time than did the basic airframe.

The new equipment brought with it new limitations on such factors as weight, maneuver and temperatures. With these restrictions came—belatedly—a new understanding that aircraft had become platforms for people and equipment, rather than simple machines composed of power plant and airframe. Designers were forced to turn their backs on the lessons of the previous seven decades, during which advances generally had been sought in new airframes and engines, not in the integration of systems.

One nation has adopted a different method of beating the high costs of aircraft research and design. The Soviets adapt slowly, moving sequentially through aircraft types by means of selective improvement, keeping enormous numbers in service during far longer periods than do Western countries.

The Soviets make technical progress "on the cheap." Their strategy was to watch Western development of F-15, F-16 and F-18 fighters. A few years later, they introduced the

Sukhoi-27 and MiG-29 fighters that had characteristics similar to NATO types. To be sure, the Soviets fall a bit behind while they gather information from aviation journals and through less scholarly intelligence techniques, but eventually captured or bartered examples become available, and capabilities are translated into aircraft. Considering the limited life span of equipment and limited aircraft production rates in the West, it is ironic that the Soviets will periodically field far more aircraft of equal or greater quality—only partially offset by the introduction of new, even more sophisticated American or NATO aircraft.

The rest of the world is coming around to the general Soviet idea that aircraft can be regarded as long-lived platforms on which new systems can be mounted. Since many basic aircraft have reached such high levels of performance, it makes sense to continue their use over extended periods by remanufacturing them "from the inside out," keeping the airframe but improving communications, avionics, propulsion and other systems. In all cases, though, our black boxes are probably better than their black boxes. The free world really shines in microcomputerization.

To test the new designs and equipment as well as the new flight regimes they made possible, experimenters built specialized aircraft designed to achieve particular test goals. The test pilot became a scientist as well as an artist. Perhaps the

Fighter tactics and missile avoidance techniques evolved rapidly in Vietnam. Outstanding fighter of the era, the big McDonnell F-4 Phantom II was originally armed only with air-to-air missiles. Above, USAF Col. (later Gen.) Daniel "Chappie" James, Jr., stands with his F-4 Phantom II at Ubon Royal Thai Air Force Base, Thailand, in 1967. James led a flight of F-4s in the 1967 Bolo MiG sweep during which seven North Vietnamese MiGs were destroyed in one mission. James flew 78 combat missions in Vietnam, 101 in Korea. Top, an Israeli F-16 fires cannon into a Syrian MiG in Robert Taylor's painting, "Bekaa Valley Gunfight." During the largest jet battle in recent history, in June 1982 over eastern Lebanon, Israeli fighters destroyed 85 Syrian aircraft while losing none themselves in air-to-air combat. An SA-2 surface-to-air missile stopped the U.S. Phantom, right, over North Vietnam. Detonation of the SA-2 warhead filled the air with lethal fragments.

Evolution of Fighter Tactics: Present

SAM Break—*(Arab/Israeli War of Attrition) The Sinai, 1970: A Soviet SA-2 surface-to-air missile (SAM) battery with the Egyptian Army (lower right) "acquires" an Israeli F-4 Phantom II on radar (upper left). Alerted to a "lock on" by his radar-warning receiver, the pilot spots the launch cloud and dives toward the SAM site. The missile reaches the top of its "boost phase" and adjusts its trajectory downward to follow the diving F-4. Gauging the missile's range, the Israeli pilot suddenly climbs at the last possible instant; unable to change its course fast enough to follow, the SA-2 explodes harmlessly.*

Although Fighting Wing and Fluid Four tactics were still used by some air forces well into the Seventies (see page 216), they have gradually been replaced by more open formations. Normally two aircraft fly side by side, separated by as little as 3,000 feet or as much as several miles. This led to tactics now called Sequential Attack, Loose Deuce, Fluid Two and others, using the two fighter team in a flexible, coordinated manner, similar to that of the two-section teams of Fluid Four. These systems have combined the inherent safety of the two-fighter team with the offensive effectiveness of two-element coordination.

The last 20 years have also seen a dramatic reversal in the trend to higher engagement altitudes set during the first 40 years of air combat. The primary reason for this is the SAM (surface-to-air missile), which has driven combat aircraft right down to the treetops and sand dunes in an attempt to exploit the weaknesses of radar at those levels.

Recent advances in radar technology, as well as such future concepts as "stealth," however, hold the promise of eventually restoring the trend to higher engagement altitudes. The flexible two-plane fighter section, though, can be expected to endure for many more years.

ROBERT L. SHAW

Evolution of Fighter Tactics: Tomorrow

A simplified description of the tactic called the Chainsaw follows—what a dogfight of the future might resemble. The two "bogeys"—unidentified hostile aircraft—and the three aircraft in the foreground are shown as multiple images.

An Airborne Warning and Control System (AWACS) E-3 Sentry over 100 miles away "vectors" three F-16 Fighting Falcons (lower left) to intercept two bogeys 20 miles distant, closing at the speed of sound. Traveling in a chain two to three miles apart, the Falcons accelerate to "Mach 1" just as the bogeys "acquire" the F-16s on radar. Thirty seconds later and 10 miles closer, one bogey achieves radar "lock" on the lead (red) Falcon and fires a radar-homing missile. Alerted by his radar-warning receiver, the red F-16 breaks hard left, dumps "chaff"—a cloud of metalized glass fiber that forms a large radar image. He also turns on his "music"—electronic countermeasures that attempt to deceive or jam radar—before he flies out of range. The white Falcon assumes the lead of the formation as the radar-homing missile loses its target and veers off. By the time the two formations are five miles apart, the bogeys know they've missed and lock on the second (white) F-16. As his radar-warning receiver sounds, the white Falcon breaks left, dumps chaff, turns on his countermeasures and "snap rolls" around the missile, which barely misses him. Simultaneously, the blue F-16 secures a good "lock" and "squeezes off" one shorter-range, "all aspect," heat-seeking missile. As the bogeys become tiny specks to the Falcon pilots, the missile destroys one enemy aircraft. The remaining bogey has little chance against the three F-16s.

Computer generated graphics flash on a transparent head-up display (HUD) at pilot eye level. Seconds become critical during high-speed maneuvers, and a HUD enables a fighter pilot to see continually updated information on weapon systems, radar guidance and flight orientation without shifting attention to the instrument panel.

most famous combination of test plane and test pilot was that of the Bell XS-1 *Glamorous Glennis* and its pilot, Charles E. "Chuck" Yeager. Together, they broke the "sound barrier" on October 14, 1947. The three XS-1s from the Bell factory in Buffalo, New York, were followed by two Bell X-2s, one of which reached Mach 3.2 but killed its pilot after going out of control. The X designation (shorthand for experimental) was no guarantee of success, though. Not only did the stilettolike Douglas X-3 never achieve its goal of Mach 2, but it also attained Mach 1 only in a dive. Puny engines were to blame. The Northrop X-4 demonstrated some of the problems but not all of the solutions of tailless flight. The Bell X-5 was essentially a modernized version of a German experimental aircraft with variable-sweep wings from World War II.

The most famous and influential of the series was the coal-black North American X-15, which made its first powered flight on September 17, 1959. Three were built, and they set records for speed and altitude as yet unexceeded by any other aircraft, excepting the space shuttles. In 1963, Joe Walker reached an altitude of 354,200 feet or about 67 miles, and four years later, Bill Knight achieved a speed of 4,534 miles an hour in the X-15A-2.

Much modified, the X-15 could have become an orbital vehicle, a true predecessor of the space shuttle—almost two decades in advance. Sadly, the program was scrapped because of time and money constraints.

Against this background of research and development in America, the specter of war once again arose. Brush-fire conflicts popped up immediately after 1945. In this same year a little-reported revolt in Iran's province of Azerbaijan signaled the beginning of the Cold War between the United States and the Soviet Union. In Korea during 1950, the Cold War turned hot.

The past met the future in Korean skies. Veteran World War II planes again performed the tactics used five years earlier. On the other hand, for the first time in history, jet fighters tangled in combat. Korea was the anvil upon which would be forged many tactics and technical advances in modern air-to-air combat.

The air war intensified when Russian-built MiG-15s (used by Communist forces) attempted to establish air superiority by sheer numbers in what became known as MiG Alley. The United Nations forces countered with various types, particularly the USAF F-86 Sabre and Navy F9F Panther, which established air dominance in battles that raged until 1953. However, despite United Nations air superiority, the Korean war ended in stalemate on the ground.

In Korea, the great powers began to learn strategic lessons, many of which involved air power. The least recognized but most momentous strategy involved the introduction of artificial, politically dictated rules of engagement. In the case of Korea, these were intended to avoid overtly offending China

Some of the world's hottest planes, McDonnell Douglas F-15 Eagles, left, stand on the wet tarmac at Langley Air Force Base in Hampton, Virginia. Mark Nowotney, below, polishes the canopy of his aircraft at the USAF's Test Pilot School. Dirty Plexiglas is anathema to combat pilots, as a "speck" on the horizon can almost instantly become a hostile aircraft in firing position. A Navy F/A-18 Hornet appears at right. After a long period of development and testing, this multimission aircraft first reached the fleet during the mid-Eighties.

and drawing her officially into a war in which she was unofficially totally immersed.

Other lessons learned from the Korean experience had a familiar ring. The United States found that the cost of the pell-mell buildup of forces to fight the Korean War and their deployment across the Pacific was ten or fifteen times more expensive than maintaining a strong military presence in the region in the first place.

In many ways, Vietnam proved more stern a schoolmaster than did the Korean conflict. From 1945 until their defeat in 1954, the French had fought a ground battle they were almost certain to lose, with totally inadequate air forces. When the French left, hostilities resumed between North and South, nurtured by other nations that gradually donated supplies, equipment, and advisers.

The United States sponsored a tremendous buildup of materiel and American fighting men from 1965 to 1972. Restricted on land, sea and air by rules of engagement dictated by both political perceptions and realities, the Americans conducted a war they were destined to lose. Perhaps the war's greatest irony was that combatants on both sides found the American rules of engagement unbelievable.

Perhaps the only means the United States had of achieving a victory in arms, the unrestricted use of air power against North Vietnam, was forbidden until the failure of other policies. While concentrating its air power in the South against the "insurgents," the U.S. sent in ground troops and worked to shape up the South Vietnamese army. Then, implementing the policy of Vietnamization—one of the worst coined words in history—the United States decided to remove its ground forces and leave the fighting to the South Vietnamese, supported by U.S. air forces. The effect was similar to sawing off two legs of a tripod.

Meanwhile, the North Vietnamese had created the most formidable system of air defense in the world, more modern and more concentrated than any faced by air forces before. When the North Vietnamese withdrew from negotiations in Paris to take advantage of the weakened condition of the South in 1972, the U.S. air forces conducted unrestricted attacks against Hanoi and Haiphong, including massive strikes by B-52 bombers and other aircraft and in a furious December campaign forced North Vietnam back to the peace table. Later, of course, in 1975 the North Vietnamese abrogated the 1972 treaty accords and swept south to Saigon.

As something of a consolation prize, the U.S. learned quite a bit about air power in the modern age of jets, missiles and highly sophisticated radar systems. First, air-to-air missiles were not enough to rid the sky of enemy jets. Fighters still

needed a gun, as many pilots had always argued. Furthermore, the new generations of computerized fighters needed to be more "user friendly." Jet combat was too complex for systems that called upon a pilot and a weapon systems officer to coordinate intricate combat in a matter of seconds. The experts determined that iron bombs—conventional explosives—could rain down almost indefinitely on the countryside with no significant result, because there were few distinct targets in South Vietnam. It was also learned that air operations conducted from 12,000 miles away were cumbersome and unresponsive to the immediate situation. And more than anything, the Vietnam experience demonstrated that the day-in and day-out maintenance of a conventional air war is too costly even for a nation as rich as the United States.

Tiny Vietnam, a sliver of a nation perched between mountain ridges and ocean, had absorbed all the ordnance that the greatest industrial nation on earth could throw at it during seven years and had shot down thousands of America's best aircraft in the process. In high-level strategic terms, the U.S. learned that a war on the scale of World War II could no longer be supported. In blunt language, this meant that any major war would immediately escalate into a nuclear engagement. How tragic that this is all any nation will be able to afford; how tragic that the "nukes" are so inexpensive compared with conventional equipment and standing armies.

Other lessons emerged from aerial combat at the opposite end of the continent, in Southwest Asia—not far from Biblical Armageddon, the prophesied site of humanity's final armed conflict. "Born in Battle" is a literal description of the Israeli Air Force. Since Israel's battle for independence in 1947, constant conflict has forged its small but excellent armed forces, establishing the tiny country as a major political power in the region. Faced with enormous strategic problems—the entire country can be swept by missiles, and key population areas are only minutes away from enemy attack—the Israeli Air Force had created an image that has itself become a shield.

In the skies over Israel and Lebanon, many modern fighters have engaged in mortal combat: MiG-21s and 23s, F-4s, F-15s, and F-16s and Dassault Mirages. Here the leading edge of the world's air forces have fought again and again. Of all the nations in the world, Israel has had to learn the most and be the best. Some lessons have been costly; it was almost overcome by the Arab introduction of huge quantities of Soviet provided SA-2, SA-3 surface-to-air missiles, the SA-6

Five U.S. jet-age veterans, from left: Doug Benefield, chief test pilot for Rockwell International; Chuck Sewell, chief test pilot for Grumman; Tony Levier, director of flight operations for Skyfox Corporation; Russell O'Quinn, chief test pilot and president of Skyfox; Darrell Cornell, chief test pilot for Northrop.

The Northrop F-5F Tiger II, above, flies for Saudi Arabia. Primarily for sale abroad, the F-5 series provided high performance at a reasonable cost. One of the latest models in the made-for-export fighter market is the F-20 Tigershark, left, a F-5 with improved avionics and a single large turbofan engine.

An A-7 Corsair II, above, guns its engine to maximum thrust seconds before a giant catapult will launch it off the carrier at upwards of 150 knots. Mobilized within minutes, Corsair IIs, F-14 Tomcats, A-6 Intruders and S-3 Vikings take to the skies from the deck of the U.S.S. Nimitz*—a formidable array, yet only a fraction of the carrier's 90-aircraft capacity.*

and SA-7 mobile missiles and the ZSU-23 multibarreled antiaircraft guns with radar control. The efficient employment of these arms by Israel's enemies forced Israeli development of an entirely new program of electronic countermeasures during the Yom Kippur war of 1973.

The tiny country's success in devising new tactics and equipment became evident in 1982, when the Israelis destroyed the Syrian missile air defense system in Lebanon's Bekaa and later destroyed the Syrian fighter force. Both times the "kill ratio" was 85 to 0 in favor of Israel. To achieve such a startling score, Remotely Piloted Vehicles (RPVs) pinpointed enemy installations and acted as pilotless drones to draw fire. Above all, the Israeli Air Force demonstrates the effectiveness of real proficiency and experience in aerial warfare.

Technical advances from warfare, spinoffs, as NASA calls them, have enriched commercial aviation and the civilian economy in general. Onboard and ground-based computers give pilots far more precise instrumentation. In addition, satellite weather prediction and better communications have become part of daily life. In other areas, such as airframe design, the public benefits have been somewhat more limited. Several military aircraft employ the swing wing, but its applicability to airlines has not been proved.

Perhaps the greatest benefit stems from the increased use of composite materials, which reduce weight. Their use directly addresses the greatest problem of contemporary civil aircraft: high purchase and operating costs. But one true revolution has occurred in general aviation: the advent of the business jet aircraft.

Before such fast, safe and capable twin-engine executive aircraft as the Aero Commander, Beech Baron or Cessna 310, flying had been more of a hobby for corporations than a tool. For years, the great Boeing Aircraft Company did without a

corporate aircraft, simply because they were not economical. Those companies that owned aircraft usually had a chief executive officer who liked to fly and who charged off the expense, either to the company or to the government.

However, when the superb Grumman Gulfstream I entered production in 1959, executive travel took a great step forward. Not only was this twin-turboprop masterpiece a Rolls-Royce of an airplane, it also made money by giving the executives it carried greater flexibility.

If the Gulfstream offered a new dimension, the executive jet revolutionized the world. Lockheed's Jetstar flew first, during September 1957, but lacked consumer appeal. It remained for William Lear to set off the revolution. Against all advice and predictions, he pushed ahead with his tiny six-seat, 500-mile-per-hour Learjet, which entered production during 1964. So small that there was not room to stand up in the aisle, based in part on a Swiss fighter design that never reached production, the glistening white Learjet took the business world by storm. It became the ultimate status symbol; even though now there are many competitors, the term "Learjet" is synonymous with executive jet.

Today, business aviation occupies an integral and honored place in air transportation. Corporate craft serve more airports and carry more passengers than do the scheduled airlines. Supersonic and other advanced designs for business jets are on the drawing boards, but existing types serve well enough for the commerce of the day.

A sampling of the latest military aircraft the world over: opposite top, the Dassault-Breguet Mirage 2000, the primary combat aircraft of the current French air force; opposite below, the newest of the U.S. Research Airplane Program, the forward sweptwing X-29, a design that promises greater maneuverability and better low-speed handling than current fighters; top, a Soviet MiG-29 (foreground), equipped with state-of-the-art avionics borrowed from the West accompanies a Tupelov Tu-26 "Backfire" bomber—one of the most versatile in the world. A U.S. Department of Defense line-up, below, compares U.S. (bottom row) and Soviet tactical aircraft. In general, the Soviets wait for the West to develop a technology, then adopt and adapt it to their own needs.

Comparable Tactical Aircraft

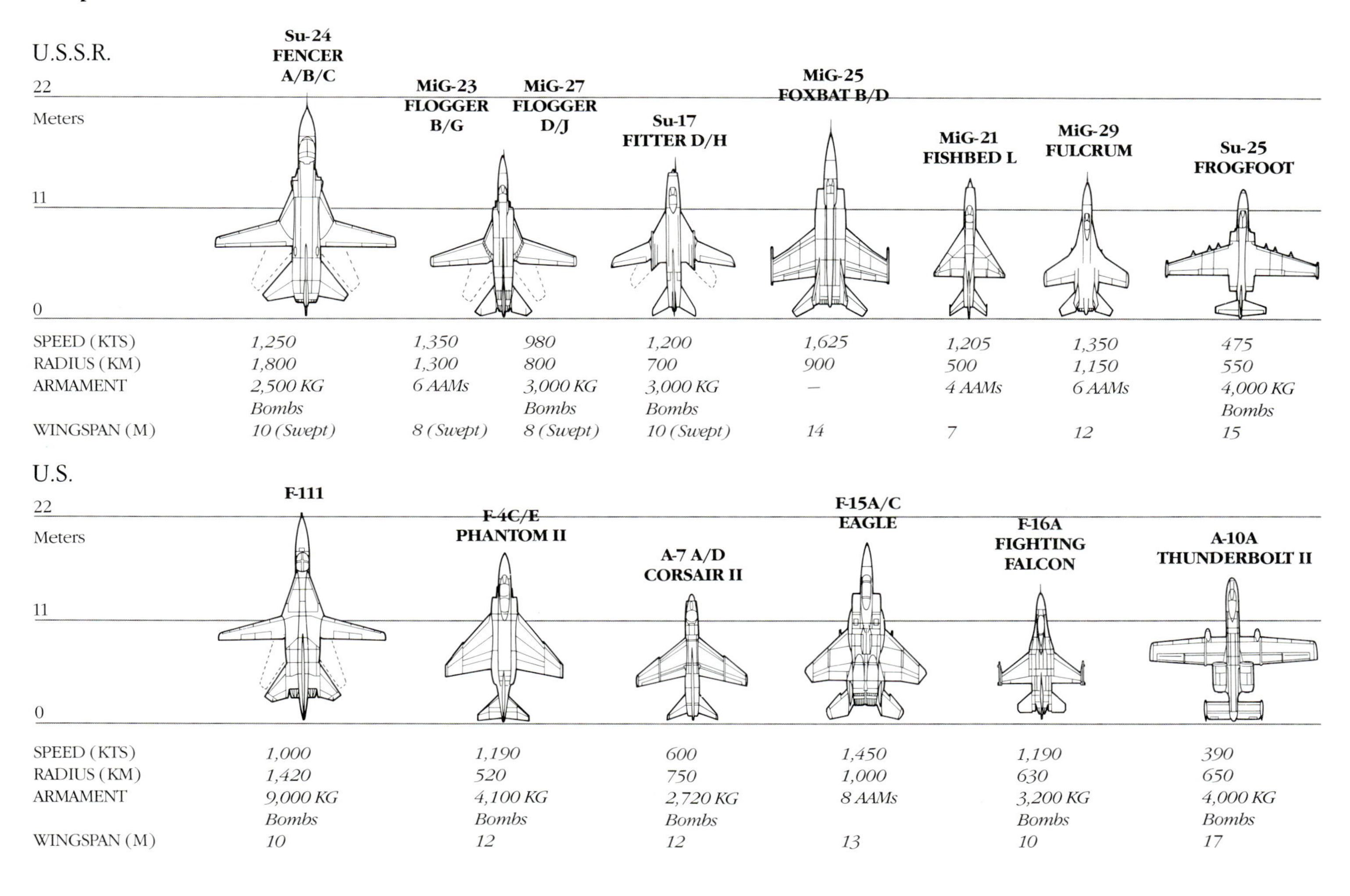

U.S.S.R.

	Su-24 FENCER A/B/C	MiG-23 FLOGGER B/G	MiG-27 FLOGGER D/J	Su-17 FITTER D/H	MiG-25 FOXBAT B/D	MiG-21 FISHBED L	MiG-29 FULCRUM	Su-25 FROGFOOT
SPEED (KTS)	*1,250*	*1,350*	*980*	*1,200*	*1,625*	*1,205*	*1,350*	*475*
RADIUS (KM)	*1,800*	*1,300*	*800*	*700*	*900*	*500*	*1,150*	*550*
ARMAMENT	*2,500 KG Bombs*	*6 AAMs*	*3,000 KG Bombs*	*3,000 KG Bombs*	—	*4 AAMs*	*6 AAMs*	*4,000 KG Bombs*
WINGSPAN (M)	*10 (Swept)*	*8 (Swept)*	*8 (Swept)*	*10 (Swept)*	*14*	*7*	*12*	*15*

U.S.

	F-111	F-4C/E PHANTOM II	A-7 A/D CORSAIR II	F-15A/C EAGLE	F-16A FIGHTING FALCON	A-10A THUNDERBOLT II
SPEED (KTS)	*1,000*	*1,190*	*600*	*1,450*	*1,190*	*390*
RADIUS (KM)	*1,420*	*520*	*750*	*1,000*	*630*	*650*
ARMAMENT	*9,000 KG Bombs*	*4,100 KG Bombs*	*2,720 KG Bombs*	*8 AAMs*	*3,200 KG Bombs*	*4,000 KG Bombs*
WINGSPAN (M)	*10*	*12*	*12*	*13*	*10*	*17*

Bridging the gap between airplane and helicopter, the British Aerospace Harrier became the first fixed wing, operational jet able to takeoff both horizontally, left, or vertically, below. The Harrier's secret is four exhaust nozzles that the pilot can rotate, changing the angle of engine thrust even while flying. One Harrier claimed the vertical takeoff time-to-altitude record, climbing to the height of Mount Everest in one minute and 44 seconds. Though tricky to operate, such craft proved their effectiveness in combat during the Falklands War.

For the scheduled airlines, the success of the 707s and 727s created an air-traffic glut by the mid-Sixties. Because bigger aircraft were always better in terms of direct operating costs (it costs 2.4 cents per air-seat mile in a 100-passenger jet compared to 1.8 cents per air-seat mile in a jet carrying 400 passengers), Juan Trippe of Pan American World Airways and William Allen of Boeing decided to take a gamble on the 747 widebody, and they won. After a rocky start, the 747 became the world's premier subsonic airliner and remains so. This profitable behemoth all but shrugs off the competition, smaller widebodies such as the McDonnell Douglas DC-10, Lockheed L.1011 or the Airbus Industries A300.

The mid-Sixties also saw the introduction of the supersonic transport (SST). America's ambitious program called for a Mach 2.7 machine, and Boeing won the long and grueling competition, only to have the program terminated for a variety of reasons, including environmental considerations. The Russian entrant in the SST field, the TU-144—dubbed the Concordsky in the West for its similarity to the British-French aircraft—was withdrawn from service in 1977, after only ten months of operation. In 1973, a TU-144 prototype had broken up in midair during a demonstration at the Paris Air Show, killing the crew and several people on the ground; the cause of the crash was never revealed.

It was Britain and France, however, that launched an era of international cooperation with the Mach 2 Concorde, smaller, slower and less economical than the proposed American craft. The high unit cost, the rise in fuel prices and the limited range and capacity of the aircraft restricted production of the

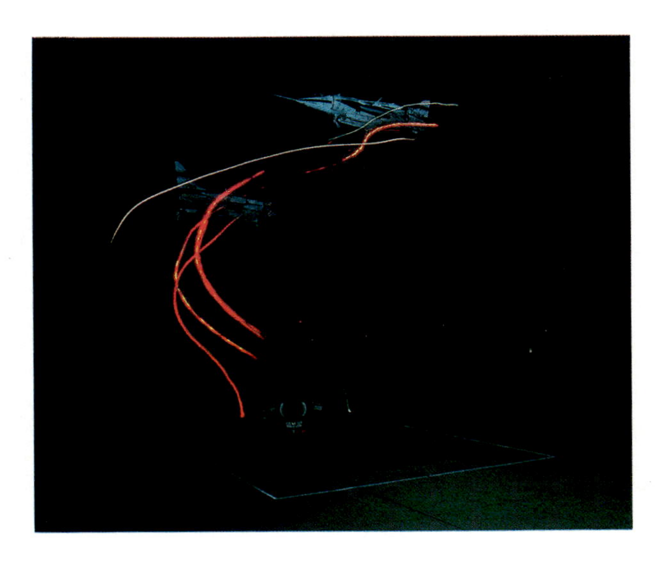

X-15 Test Pilot

A. Scott Crossfield

The X-15 flew for ten years, some 200 flights, with 12 pilots. With its brute strength and marvelous controllability it gathered more data in unknown regions than any airplane in history. Rocket powered, it set records for speed and altitude which have never been even approached by any other airplane.

The X-15 flew very well, responded as a fine mount should. There was little that a pilot couldn't do if he flew with skill and attention and avoided abuse. But there were a few things one just could not do and still stay in the saddle. I would never attempt a deep stall with the X-15 or waste my time trying to recover from a spin should I have been so imprudent as to get into one.

Its fine breeding showed on the third flight, when one of the LR-11 rocket chambers blew up at light-off. The exploding rocket chamber blew out the aft bulkhead, designed to give way in such an event, caught fire, snuffed itself out, and then (since I did a few of the right things) the badly wounded X-15 brought me down safely.

Unlike ships and most airplanes the X-15 was never referred to in the feminine gender. No one ever referred to the X-15 as "her" or "she." The X-15 was an eager ready stallion all the way. We simply referred to it as the X-15. The news media, of course, gave it their usual trendy swaggering hip treatment: the "Blackbird," the "Manned Missile," and sometimes with apparent glee, the "trouble plagued X-15," and the like.

All of the research airplanes had what might be called personality traits in their flying and handling qualities. The X-1 was a jewel, very easy to maneuver, very stable and forgiving. The X-4 controls had a characteristic sensitivity and power that would give the pilot a way out of its many stability tantrums born of that sensitivity. It would bite you when you weren't looking. The X-5 had no personality in its flying qualities at all and some bad behavior in its instabilities. The XF-92 was a "dog," period. The D-558-1 was similar to the X-1 but wasn't fun to fly like the X-1 was. The D-558-2 took a delicate touch to bring out all of its best characteristics and to avoid its idiosyncrasies. A heavy hand could bring on trouble.

The X-15 had to be flown attentively and intelligently. For what it was asked to do, however, the X-15 looked after its pilots very well. It was an aviator's airplane in the finest sense. As of the day of this writing, an airplane to match its performance and handling qualities has yet to be built.

Test pilot Scott Crossfield prepares for the first captive flight test of the North American X-15 rocket research aircraft, top, in 1959. Subsequent X-15 flights established the still unbroken world's absolute speed record and the absolute altitude record. Each of the nearly 200 test flights began with a drop launch from a B-52 Stratofortress, above.

The space shuttle, left, with its airplane shape—and huge, disposable booster rockets—is part aircraft, part missile. Below, future military aircraft may resemble the low radar signature "stealth" aircraft, depicted here in an artist's concept.

Concorde to 16 aircraft. This token entry assuaged the national pride of Britain and France, though research and development costs could never be recovered. Having established a niche for itself with wealthy and business travelers the Concorde celebrated its tenth year of service during 1985, still evoking prestige, champagne and miraculous speed.

The selfsame fuel rise that thwarted the mass production of the Concorde caused a revolution in thinking about the economics of new transports. Operating economy became synonymous with fuel economy. New fuel-efficient engines and lightweight materials in the Boeing 757, 767 and Airbus 310s drove down seat-mile costs, creating a trend that may extend into the next century. By then, alternate fuels may permit the development of the hypersonic airliner, called variously the National Space Plane, the Orient Express or the Trans-Atmospheric Vehicle. This, "most complicated vehicle ever built" probably would be able to take off from a standard runway. Perhaps as many as four types of engines would power the aircraft: a powerful gas turbine for takeoff, a ramjet for acceleration to Mach 6, with a "scram-jet" conversion capability for acceleration to Mach 20, and finally a rocket motor to boost the vehicle halfway around the world in space at near-orbital velocity.

The most optimistic backers of the National Space Plane believe a prototype will fly by the mid-Nineties, and that a fleet will do to today's jumbo jets what the original jet airliners did to ocean liners. Many questions of materials, design and human factors remain, but with administration backing—heightened by foreign competition in the form of the Super Concorde and the British HOTOL (a slightly less ambitious ground-to-space craft)—the incentive to build the Orient Express will be enormous. Don't bet against a new China Clipper, operating between Los Angeles and Beijing in two hours.

In contrast to the startling leaps in speed and altitude of fixed-wing aircraft, helicopters—vertical flight, rotary-wing

New designs blend elements of the helicopter with those of fixed-wing craft: right, a NASA inspired design—the Bell XV-15—can tilt its rotors forward and fly like an airplane after vertical takeoff. At left, the Edgley EA7 Optica combines an aft-mounted, ducted-fan propeller with an up-front bubble canopy for splendid visibility. This slow flying, high-tech "fly" may find application in police work, forestry, engineering and aerial photography. Bottom right, scimitarlike propellers of the Unducted Fan could represent the shape of the future. Made of light, extraordinarily strong composite materials and powered by a jet engine, these thin, wide blades can propel a passenger transport up to Mach 0.8 at a substantial savings in fuel over a conventional turbojet.

Overleaf: the shape of future corporate aircraft may resemble the Beech Starship, a pusher turboprop constructed mostly of light, extremely strong composite materials instead of aluminum.

machines—have advanced more incrementally. Burdened by the great mass of the rotors and the astonishing complexity of their transmissions and control systems, helicopters can go only so fast and so high, no matter how much power they have. Yet the jet engine was an enormous breakthrough, helping to reduce vibration and provide better performance at higher altitudes. Size rather than high speed and altitude performance characterize many of today's machines, such as the Sikorsky CH-54 and Soviet MIL Mi-10 flying cranes.

Vietnam helped to maintain the helicopter's search and rescue role, with Sikorsky HH-53C "Jolly Green Giants" venturing far behind enemy lines to pick up downed crew members. But the war's most significant advance came with the introduction of the attack helicopter, primarily the Bell HueyCobra, which started a trend in army support aviation in all armies. Best known of today's generation of attack helicopters is the hulking Soviet Mi-24, the "Hind" of western military nickname makers, much used by the Soviets in Afghanistan. Here, both sides have learned that rapid deployment of army units is possible at a moment's notice with helicopters.

Throughout the history of helicopters, inventors have sought solutions to vertical flight that do not involve rotors. Suggestions have included tilt engines, tilt wings, tilt rotors and combinations of all of these, but none were really successful until the Bell XV-15. A remarkable aircraft with a 20-year, multimillion-dollar development history, the design of the Eighties at long last combines helicopter and aircraft performance into one "hot ship." The tilt-rotor technique may finally fill America's need for air transport between one city center and another.

With skyrocketing sophistication and cost, one might think of aviation as moving away from the common man. And indeed in a traditional sense it is: certainly the major manufacturers no longer address the problems of general aviation. But, as it always has, commerce has a way of finding and filling old and new economic niches with specialized products for specialized uses.

The homebuilt movement provides another means of meeting America's demand for sport and working airplanes for the general market. Homebuilders use novel construction

First of the jumbo jets, the Boeing 747, right, is the world's largest civil aircraft, capable of carrying more than 500 passengers. The dedicated work of air traffic controllers such as those at Chicago's O'Hare Airport, below, help make commercial flying one of the safest means of transportation today.

techniques and materials to create innovative aircraft. Already svelte aircraft with remarkable performance appear in profusion—landplanes, amphibians, helicopters—all from the fertile minds of homebuilders.

The ultralight airplane market took off after people began putting tiny engines on their hang gliders. Since then, the evolution of the ultralight has recapitulated the evolution of the airplane; each generation has become larger, faster and more complex. More sophisticated than an ultralight, but smaller and less expensive than the lightplane of old, these air recreational vehicles are growing up, assuming many shapes and sizes: kits that replicate the magnificent machines of the past, mini-Cubs and mini-Champs; sleek amphibians with engines mounted on pylons; stagger-wing biplanes that combine classic looks with low cost. The homebuilt aircraft phenomenon has certainly attracted the lion's share of American ingenuity—a precious cultural commodity.

Another market exists, one not as innovative as the homebuilt, but important nonetheless: the older but well maintained aircraft. In America, lightplanes are often treated as automobiles were in Britain—things to own, not necessarily to use. Carefully tended, always stored, lightplanes of the Forties and Fifties are as useful as when new. In addition, they bear the priceless patina that clings to precious antiques. Market prices suggest that the supply is adequate; except for exotica such as the Luscombe Phantom or Spartan Executive, prices have climbed only slowly across the years.

The imperative to fly is constant, and no matter what the intrusions of inflation, technology or regulations, men and women *will* find a way to become airborne.

ORIENT

Flight to the Future

Man first dreamed of flight not as a means to conquer the world, nor to fly from coast to coast or to make money—it began as an urge simply to be liberated from the ground, to see the earth from another perspective and to rendezvous aloft with the birds and the winds. In a sense, this dream is eternal, thriving in our own times, and for the common man it could be realized in the first half of the next century. That, of course, will be flight's real golden age, a time worthy of aviation's remarkable past.

In a sense, the marvelous history of flight makes our time part of the possibilities of tomorrow. Today, enthusiasts search the proven technologies as they design and build new experimental planes. And oddly enough, in the current recapitulation of the development of aviation, we see the reemergence of aircraft similar to those of long ago—fun to fly

The world watched raptly as pilots Richard G. Rutan and Jeana Yeager flew Voyager, *left, around the world without stopping or refueling during nine grueling days from December 14 to 23, 1986. Designed by Rutan's brother, Burt, and built of advanced composite materials, the 111-foot-wingspan craft (which weighs 2,680 pounds empty) took on 9,000 pounds of gasoline for the flight.* Voyager's *aerial circumnavigation of the globe on one tank of gas captured the world's imagination.*

In a delightful anachronism, artist Guy Johnson evokes the great eras of air shows and races, below. At the 1911 Circuit of Europe, Jules Védrines's Deperdussin appears ready to race James "Jimmy" Doolittle in his formidable Gee Bee of 1932, the Thompson Trophy winner. Left, a tandem-wing hang glider designed by California gliding pioneer John Montgomery descends during a 1905 San Jose air show. "Tiny" Broadwick, first woman to parachute from an airplane, poses with pilot Glenn L. Martin, opposite, below. At top, the de Bothezat helicopter hovers. In 1922 it flew for 102 seconds in a U.S. Army test, but funding was not forthcoming.

if you are willing to invest the time and energy in learning how to do it. In fact, in some ways we seem to be backing into our future.

In terms of genuine personal use of aircraft we are currently not far past the Kitty Hawk stage of flight. For the past eight decades, all of the progress that has been made in aviation has been in the direction begun by the Wright brothers. Only in the next decade, because of the advances made in other sciences—particularly in computers and in materials—are we going to reach a point where aviation will be translated directly to individual use at the same or greater degree of an automobile or bicycle. And in this seemingly impossible and withal innocuous happening are involved philosophical and political considerations which have great import for all mankind.

While the public has always demanded a "universal flyer," an aircraft that "anyone can fly," there has never been a serious attempt to achieve this goal. Certainly there have been small, easy-to-fly airplanes, from the perilous Pou de Ciel to the Bede 5J: but even the products of major manufacturers—Piper Cubs, Aeroncas and the like—have all demanded more than casual skill and dedication.

While our involvement with the past can be very constructive, we must look ahead. We would do well to set our sights high, and we already have some evidence that this is happening. We have seen some remarkable approaches to the dream of Everyman's flight in the *Gossamer Condor* and other human-powered vehicles built by Californian Paul MacCready. These conquered the Kremer Prize for human-powered flight one year and the English Channel the next.

Other innovators include a team of scientists at the Massachusetts Institute of Technology. They are working on a 110-foot-wingspan human-powered aircraft with a gross weight of less than 70 pounds. This aircraft is being designed to

LOCKHEED
"VEGA"

The story of flight is full of stunning accomplishments and some not so stunning. Five years to the day after Charles Lindbergh's solo transatlantic flight, Amelia Earhart, left, did it again in her Lockheed Vega, thereby becoming the first woman to fly the Atlantic alone. Above, despite great ingenuity and the best of intentions, some machines such as this five-winged craft built by Jerome S. Zerbe in 1910, were doomed to failure.

complete the flight attempted by Daedalus and Icarus of mythic times, from Crete to the Grecian mainland, a distance of almost 90 miles.

A number of solar-powered craft draw their energy from sunlight and convert it directly into propulsion. Other aircraft designs feature computers that can take data from a dozen sources and integrate it into control inputs that maintain equilibrium in flight. Recently, the potential of computers in aviation was demonstrated in the flights of the mechanical representation of a giant Cretaceous pterosaur, *Quetzalcoatlus northropi,* created by MacCready's team for the Smithsonian's National Air and Space Museum. If this unstable vehicle, with its tailless configuration and requirement for wing flapping, could be made to fly with the help of on-board computers, what then could be done with an aircraft designed to be inherently stable?

In short, what man has always wanted is the *gift* of wings, a vehicle that would lift its pilot easily and safely from the ground to the sky. What is needed is an airplane that an individual can learn to master with the same ease and relative safety that a person has when he or she learns to operate a bicycle, motor scooter or small sailboat. And now, at last, we may soon witness the creation of such a machine.

Imagine an aircraft of some 25-foot wingspan, powered by one or two very small engines and equipped with computers and sensors to give it automatic stability and control. Mass manufactured of modern composite materials, light in weight

With large planes or small, aviation buffs, both professional and hobbyist, delight in aerial craftsmanship. Above, a modeler readies his miniature Mirage jet fighter for takeoff. Opposite, Richard Horigan and Dave Peterson (at right) study the Enola Gay, *the B-29 Superfortress used to drop an atomic bomb on Hiroshima in World War II. They helped preserve this fateful ship, stored at the National Air and Space Museum's Garber Facility. Here, craftsmen painstakingly restored the Albatros D. Va World War I German fighter, right, from its badly deteriorated condition. The aircraft is one of two of the type left in the world.*

and designed to act as a resilient cushion in the event of a collision, such an airplane would be able to fly at its maximum speed into a brick wall without serious injury to the passengers. It is not here yet, but it is certainly possible.

For the most part, though, collisions in our miraculous brainchild—call it the "Safebird"—would be avoidable. Given the capability of modern sensing devices and computers to transmit and receive and to indicate or even perform avoidance manuevers, this technology could be achieved and perhaps commercialized early in the next century.

Our once-and-future ship would not be supersonic—far from it. It would be capable of the speed of bird flight, perhaps 15 to 70 miles per hour. The pilot of such an aircraft would be able to climb to a few hundred feet, circle silently with the hawks, savoring the sky and taking in the view in safety far beyond that possible on the highway—a misnomer if there ever was one.

Before we become too skeptical about this somewhat highblown concept, we should recognize that we are at the very beginning of the revolution in computers, in materials, even in aerodynamics. Once laboriously wrung from slide rules, new airfoil designs can now be turned out by the thousands through computer analysis. Each new material seems to lead to others, and lighter, quieter, more efficient power plants and propellers will certainly appear.

The achievement of the Safebird is but a matter of time—far less time than the casual reader might imagine. Of course the same improvements will make possible new standards of performance in executive, transport and military aircraft. Then, before you laugh and say that people will not put their lives in the hands of microchip computers, remember that it is done every moment of every day. Heart patients rely on pacemakers, aircrews navigate with inertial navigation systems, hospital personnel monitor the critically ill with sophisticated medical equipment and the major powers depend on computer-controlled systems to prevent the accidental use of nuclear weapons. Why not use such technology to expand our flying opportunities.

ENOLA
GAY
82
PLEASE DO NOT TOUCH

The days of aircraft design by cut-and-try methods are long past. Engineers use powerful research tools to predict the performance of new aircraft. Tiny bubbles in a NASA water tunnel show vortices in the wake of a test surface, below. British and French engineers used dyes injected in a fluid flow around a model, opposite, to show the aerodynamics of the Mach 2 Concorde Supersonic Transport.

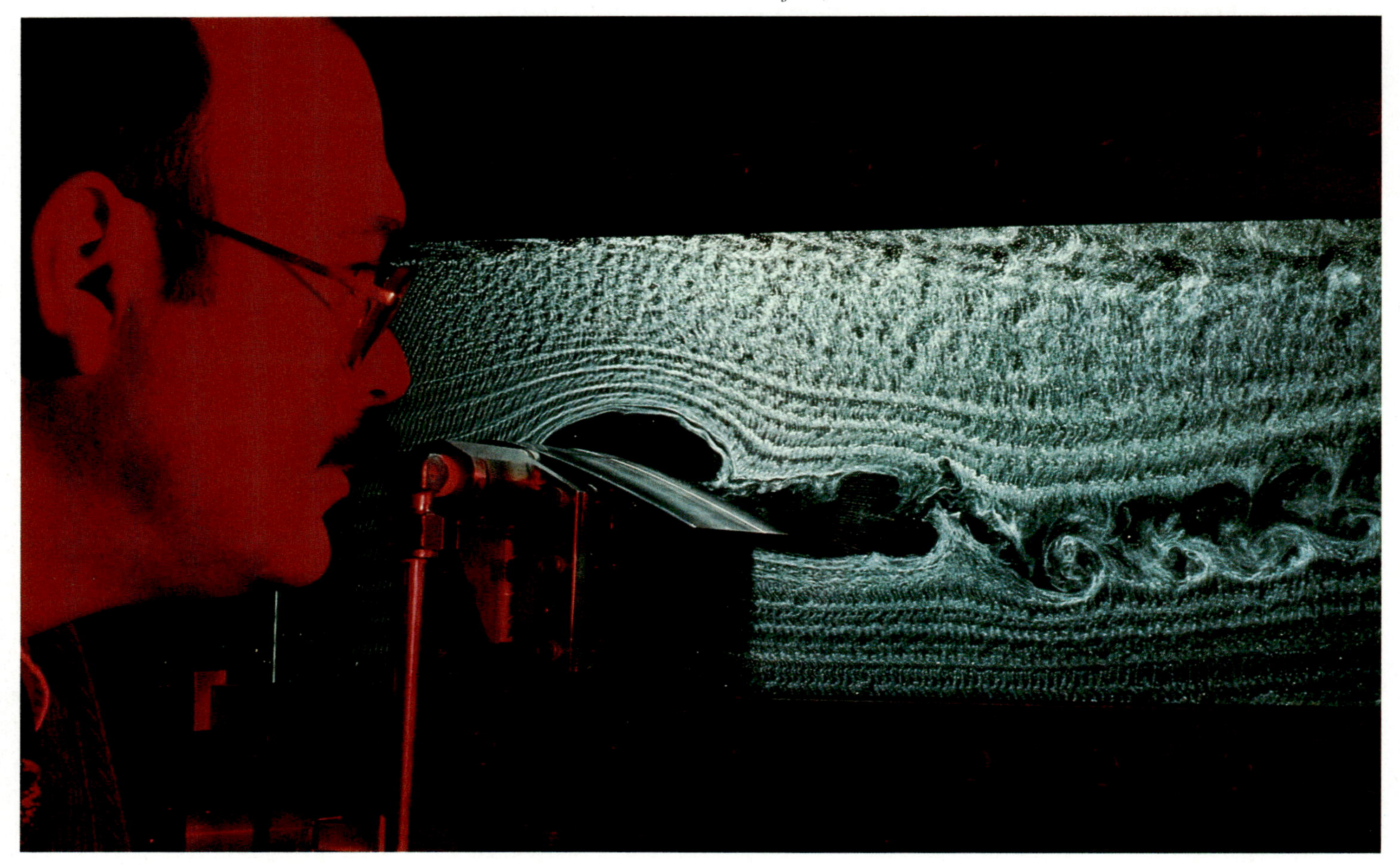

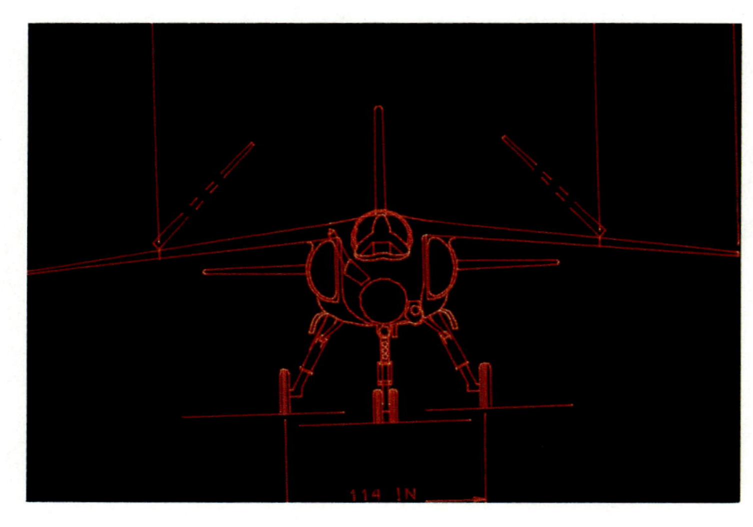

Essential in modern aircraft design, computers perform the millions of calculations needed to solve problems in stress analysis, aerodynamics and other areas. Increasingly, computers are used to reveal various aspects of aircraft design, as above.

More important from a strategic sense (if not as much fun) are flight possibilities at the other end of the size and speed scale. Where air and space merge, the future offers advancement on a grand scale. The National Space Plane, far from being a visionary extravagance, will become as familiar and as reliable as a freight train, uniting the world as the transcontinental railroad united our nation. For trips to orbiting space stations and lunar or Martian colonies, space travel will become routine for the specialists who need to go there.

Scoffers will immediately cite costs as overwhelming obstacles to all of this, but they will not have thought the problem through. In a world in which no city is farther than two hours from another, one expenditure can be reduced to the requirements of local police forces—the drain of national defense.

No single nation will be able to afford the efforts required to establish permanent space stations and permanent colonies on Mars; therefore there must be joint efforts to do so. If the 21st century's superpower contenders—China, Japan, the United States, the Soviet Union, a European Union, the South American bloc—engage in a joint technological effort

Wings for Fun: Tomorrow's Light Plane

Paul B. MacCready

Fifty years ago the typical features of the light aircraft—those planes that carry from one to five people—were already established: a single engine in the nose, a horizontal stabilizer in the rear and a skin of fabric or sometimes metal. Myriads of more exotic configurations have been developed, including autogyros and helicopters, but most light aircraft flown today are of this typical design.

A variety of factors besides basic vehicle design influences the light plane industry: pilot certification, insurance, air traffic control, instruments, availability of weather information, airports, training and vehicle sales and servicing operations are all part of the light aircraft system. Many of these nontechnical factors will have an inhibiting effect on the full use of modern and future technology. Inevitably, airspace limitations will delay and restrict some private flying and require a greater degree of pilot training for long-distance flights. Fully automated airplane operation has now reduced the need for piloting skill, but not the need for judgment in radio techniques and weather interpretations. Product liability, a general insurance problem of our culture, looms especially large.

The success of any given design (as judged by its sales) requires consideration of all these factors plus the intangibles of style and public taste, the presumption of safety and, of course, the consideration of cost.

Aircraft are used for recreation, personal transportation and business (besides transportation, there is surveillance, photography, crop dusting, banner towing and cargo carrying, for instance). For most of these tasks, cars and airlines provide competition

Entirely powered by electricity from solar cells, the Gossamer Penguin, *above, takes off at Shafter Field in California. Paul MacCready's 13-year-old son, Marshall, pilots the craft—a three-quarter scale version of another MacCready invention, the* Gossamer Albatross*—the first human-powered aircraft to cross the English Channel. Yet another MacCready development, above right, is a pterosaur, a wing-flapping, computer-assisted model of a winged reptile that lived 70 million years ago, during the Age of Dinosaurs.*

that will steadily become more formidable. The biggest opportunities for the future seem to be in recreation, personal transportation for short distances and business. Thus, there will be increasing emphasis on the low speed portions of the flight: the landing and takeoff maneuvers. There will also be emphasis on safety and comfort and, of course, on convenience and economy.

Over the past 40 years, one class of light aircraft has made full use of technological improvements—sailplanes (especially European designs). Carbon and glass composite vehicles, with the latest in airfoils and boundary layer management, now reach glide ratios exceeding 60:1 (60 feet forward for every foot of descent) and use sophisticated instrumentation and onboard computers for flight optimization.

Experimental aircraft, out of the mainstream of certificated production, have been midway between sailplanes and ordinary lightplanes in their application of advanced aerodynamics and structures.

Here are a few of the many technologies that will change the lightplane industry in the years to come:

Structure. Composites, (especially carbon, Kevlar and glass with foam or honeycomb plastics) permit aerodynamically elegant shapes of great strength and smoothness. High aspect ratio (long, thin) wings also become feasible (witness modern sailplanes and *Voyager*). Wings can be so light

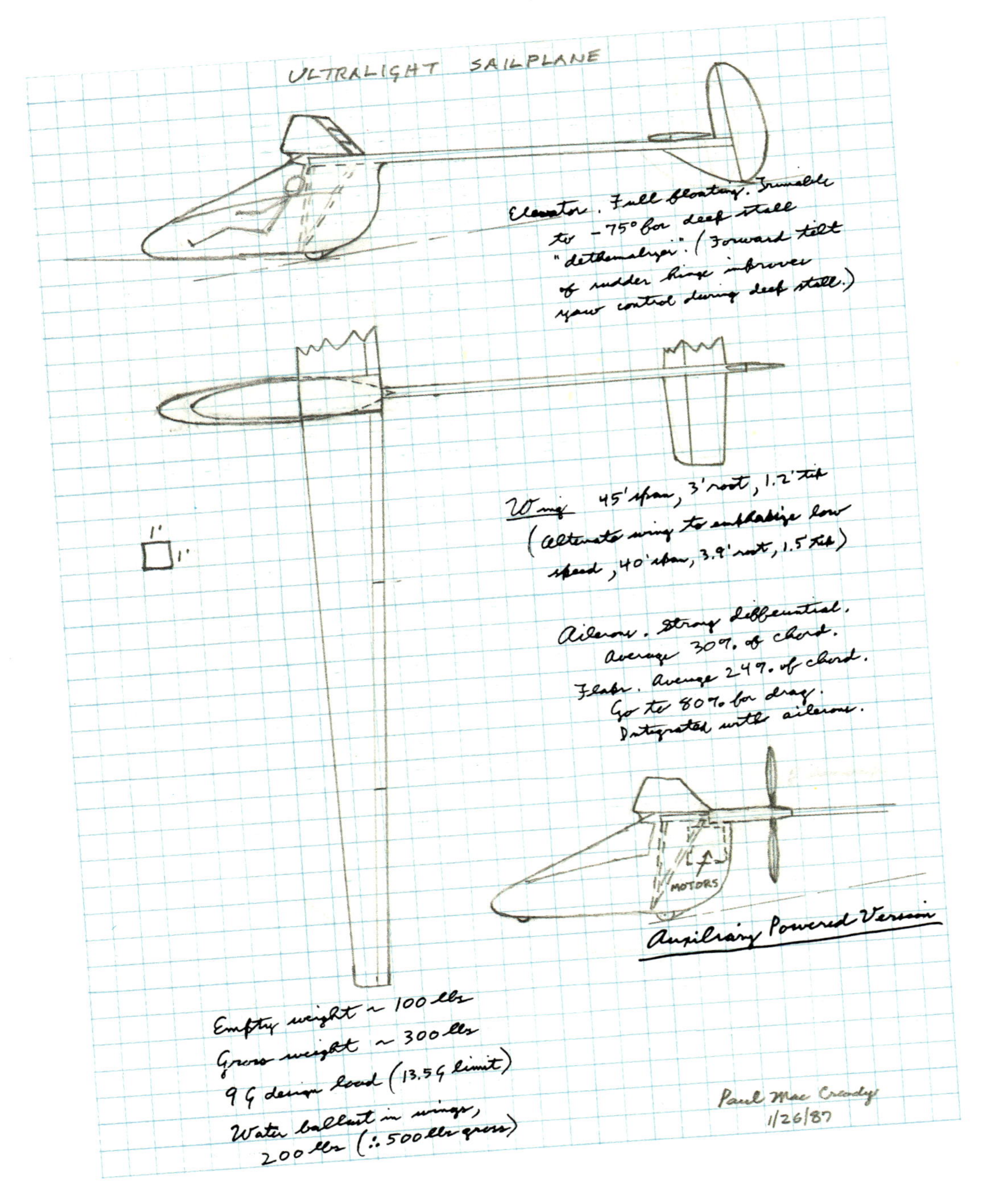

Author Walter Boyne and the editors of Smithsonian Books asked Paul MacCready to develop some concepts for an aircraft like the "Safebird" described in this chapter. Such a light plane, with its sophisticated materials and technology, would enable almost anyone to fly safely, even people with little experience. Peering into the next decade-and-a-half, MacCready suggests a design somewhat akin to today's powered soaring airplanes with their high aspect ratio (long, narrow) wings. Among the components are strong but lightweight composites for the airframe, two small auxiliary engines for powered flight and several onboard computers that constantly adjust the control surfaces of wings and tail. The automated features would help prevent stalls and other potentially fatal conditions caused by pilot error and would make the aircraft far easier to fly than other light planes.

while having the required strength and stiffness that there seems no need for ingenious structures such as joined, multiple wings.

Aerodynamics. Aerodynamic design is now so advanced that the performance and stability and control of a small airplane can be predicted quite accurately from theory. Parasite drag of smooth shapes can be minimized. The new opportunities arise for strongly distorted flows: airfoils at extremely high lifts, using flaps and/or vortex generators, fat airfoils employing boundary layer suction or blowing, stubby fuselages with low-drag aft portions to name a few. As aircraft become more streamlined and faster, greater benefit will result from careful design of details: antennas, cooling, flap hinges and landing gear doors.

Safety/Survivability. A major advance is just now reaching the experimental stage in the light aircraft field: the ballistically deployed parachute that permits the entire airplane to descend to a reasonably gentle crash landing after an emergency (loss of control, structural failure, midair collision, poor visibility, inadequate landing site). Encouraging tests are being made on the use of a gliding chute that permits some control to be maintained during the descent. Research continues on techniques to increase crashworthiness through cockpit design with composite structures.

Power. The basic, reliable four-cycle

engine will continue to be a main power source. Improvements will be in reduced cost, increasing the time between overhaul, power-to-weight ratio and in fuel consumption efficiency. Liquid cooling improves reliability and longevity and cuts down air-cooling drag. Ceramic engines, with better fuel economy and power-to-weight ratios, are in the offing. Gear reductions for driving propellers at lower revolutions per minute will steadily improve. Controllable pitch propellers will become more widely used, helping to expand the flight envelopes of light aircraft. Mufflers will be improved significantly as a result of pressure from both pilots and the public. Turbocharging will continue. Experiments with the rotary Wankel engine are ongoing, but it is doubtful that this configuration will have a significant impact over the next two decades. Turboprops will be making major strides, but are likely to remain too expensive for aircraft that are used only a few hundred hours a year. Similarly, jet engines will probably not be widely used for lightplanes in the foreseeable future.

Automatic Control. Autopilots will be more widely used, as will electric or hydraulic actuators of control surfaces. The big application for automatic control is in computer assisted, rapid active command of flight control surfaces and systems to meet moment-to-moment flight requirements. With proper flow-sensing as input, active control can permit an airfoil to operate close to stall without stalling, at slow speed in turbulence, and active control can serve to alleviate gust loads at high speed in turbulence. Thus the flight envelope is widened, to take better advantage of the high power, low drag and low wing loading of modern composite vehicles.

Variable Geometry. Given enough time and money, and a willingness to accept complexity and the associated troubles, convertible aircraft can be developed that are well adapted to both the low speed (takeoff and landing) portions of the flight and the high speed cruise portions. However, for light aircraft, the economic realities associated with complexity dictate that "convertiplanes" will not be common. One major exception is the "roadable" airplane. The few attempts in the past have "worked" but foundered on the problem that the compromise vehicle is a poor airplane and a poor automobile. Now, with the benefits of present composite structure and anticipated light, powerful engines, the roadable airplane may come into its own. The conveniences could overcome performance and cost disadvantages.

While we cannot see very far into the future, we can be reasonably confident that the advanced lightplane of tomorrow will display some combination of these features.

New concepts emerge from the notebook of aeronautical engineer and designer, Paul MacCready. He works at his Aero Vironment company, located in California's Simi Valley.

Overleaf: artist and model builder Attila Hejja worked with the author and MacCready to evolve this concept of a "Safebird." Paul MacCready, portrayed at the controls, flies with shore birds.

Today's hang glider emulates the birds, though its wings are only cloth-and-aluminum. Tomorrow's unmanned "solar sail" may waft itself through our planetary system, pushed by the pressure of light and pulled by the force of gravity.

to create a space colony—as inevitably they will—then the monies now spent in endless streams for "national defense" will be available for application to the space effort.

It is not so far-fetched to say that the single best hope for the avoidance of a nuclear holocaust lies in the global effort to colonize and exploit the resources of space. Perhaps only through such a grand pioneering effort will we achieve eventual peace. It is not an impossible dream that the bicentennial celebration of the Wrights' first flight will recognize that the things they sought—the liberation of man from the earth and the creation of a vehicle that would make war so terrible that universal peace was inevitable—had in fact been set on the high road to reality back at Kitty Hawk on that December morning of 1903. Though a whole host of accompanying disciplines—new mathematics, computers, materials, philosophy and politics—had to evolve, that first flight of 120 feet was the beginning of what may be the only course to save the world from itself.

Even with all our wondrous and bewildering technology, we cannot have it all. The world may very well have the choice of only one of two paths, nuclear annihilation or joint action in the exploration of outer space. The philosophy that could make the latter option possible is the same one which can—and will—at last make flight a universal possibility for all.

Index

Numbers in italic signify picture references.

U

V

W

X Y Z

The author:

Walter J. Boyne is the former director of the National Air and Space Museum. He enlisted as a private in the U.S. Air Force in 1951 and retired in 1974 with the rank of colonel and more than 5,000 hours in a variety of aircraft, including the B-50, B-47 and B-52. He has written more than 300 articles on aviation subjects and 11 books, including *Boeing B-52: A Documentary History*, *The Jet Age*, *Messerschmitt Me 262: Arrow to the Future*, *The Leading Edge*, and, with Steven L. Thompson, the recent best-selling novel *The Wild Blue*. In addition to the text in this volume, Mr. Boyne contributed "B-47 'The One to Fly'" on page 212.

Contributors:

A. Raymond Brooks wrote "Flying the SPAD XIII," found on pages 108–109. He shot down six German aircraft during World War I. His SPAD *Smith IV* is in the World War I Hall of the Smithsonian's National Air and Space Museum. Today, he lives in Summit, New Jersey and is active as a lecturer.

A. Scott Crossfield contributed "Jets, Rockets and Realism" on pages 198–201 and "X-15 Test Pilot" on page 254. He is one of the world's foremost test pilots and has flown many aircraft in the Research Airplane Program including the X-15. Currently, he is a technical advisor for the Committee on Science, Space and Technology of the U.S. House of Representatives.

Tom D. Crouch wrote "From Dream to Reality" on pages 28–31. He is author of *The Eagle Aloft: Two Centuries of the Balloon in America* (Smithsonian Institution Press, 1983) and *Seven Hawthorne Street: The Life of Wilbur and Orville Wright* (W.W. Norton, scheduled for 1989). He serves as Curator/Supervisor at the National Museum of American History's Division of Engineering and Industry at the Smithsonian.

Paul E. Garber shares a childhood memory with us in "The Wright Military 'Flyer' at Fort Myer, 1909" on pages 56–57. He is Historian Emeritus and the Ramsey Fellow at the Department of Aeronautics, of the Smithsonian's National Air and Space Museum. See the Dedication (pp. 14–15) for more information.

Paul B. MacCready developed the technical summary, "Wings for Fun: Tomorrow's Lightplane," found on pages 272–275. He is an aeronautical engineer, designer and Chairman of the Board of Aero Vironment, a company he founded. His highly successful, human-powered *Gossamer Condor* hangs in the National Air and Space Museum. Other projects include work on his second wing-flapping, computerized pterosaur model—this one with a 25-foot wingspan.

Edwards Park provides two reminiscences in "Two Wars and Between" on pages 72–77 and "The Jug in New Guinea" on page 178. "Ted" Park was a founding editor of *Smithsonian* magazine. He has written *Treasures of the Smithsonian* (Smithsonian Books, 1983) and *Nanette: An Exaggeration* (W.W. Norton, 1977), a book describing his World War II experiences as a fighter pilot in New Guinea. Currently, he writes on a variety of subjects, and his popular column, "Around the Mall and Beyond," appears in *Smithsonian* magazine.

Robert L. Shaw has provided commentary to accompany several chapters. His "Evolution of Fighter Tactics" appears on pages 98, 152, 216, 234, 241. He was a commander in the U.S. Navy Reserve, accumulating more than 4,000 hours of flight time, mostly in F-4s and F-14s. He is author of *Fighter Combat: Tactics and Maneuvering* (Naval Institute Press, 1985) and is working on another book on the history of fighter tactics for the same publisher. Currently, he is a major in the U.S. Air Force Reserve.

Richard T. Whitcomb wrote "The Area Rule for Supersonic Fighters" on page 222. He was an aeronautical research engineer at the Langley Research Center when he discovered the revolutionary "area rule" principle. Today, he is a Distinguished Research Scientist at Langley.

Historic Balloons and Airships
(See pages 34–37)

Up and away, lighter-than-air ships that made history: multiple dates show service life, while single dates are those of historic importance to particular craft and not necessarily their life; page 34 (clockwise from lower left): Montgolfier-unmanned (France 1782), Montgolfier-unmanned (France 1783), Montgolfier-Versailles (France 1783), LZ-1 (Germany 1900), LZ-24/L-3 (Germany 1914–1915), LZ-13 *Hansa* (Germany 1912–1916); page 35 (clockwise from lower left): Charliere-unmanned (France 1783), Montgolfier-de Rozier et Laurent (France 1783), LZ-13 *Hansa* (Germany 1912–1916), LZ-24/L-3 (Germany 1914–1915), North Sea Class Airship (Great Britain 1918–1919), LZ-126 Los Angeles (United States 1924–1932), N-1 *Norge* (Norway/Italy/United States 1926), *Royal Vauxhall/Nassau*-version one (Great Britain 1836), R34 (1919–1921), Blanchard and Jefferies Channel Balloon (France 1785), Charliere-Charles et Robert (France 1783); page 36 (clockwise from lower left): Giffard's Steam Airship (France 1852), *Royal Vauxhall/Nassau*-version two (Great Britain 1836), R101 (Great Britain 1929–1930), ZRS-4 *Akron*-and one of its Curtiss Sparrowhawk aircraft (United States 1931–1933), ZR-1 *Shenandoah* (United States 1923–1925), Santos-Dumont No. 6 (France 1901), *Oren* (Sweden 1897); page 37 (clockwise from lower left): *Pilgrim* (United States 1925), LZ-127 *Graf Zeppelin* (Germany 1928–1940), LZ-129 *Hindenburg* (Germany 1936–1937), *Caquot* (multiple countries 1914–1945), *Fugo* bomb balloon (Japan 1944–1945), ZPN-1 (United States 1951–1957), *Double Eagle II* (United States 1978), a selection of four Raven Sport Balloons (United States 1986); upper right corner): *Explorer* (United States 1935).

Acknowledgments

Alfred Bachmeier, *Chief, Collections Management, Preservation, Restoration and Storage Division*
Jerry Benitez, *Stanford Paper Company*
Fred Bigio, *inker*
Chris Bishop, *Aerospace Publishing Ltd.*
Steve Blackwood, *Watermark, Inc.*
John Bowen, *Colotone, Inc.*
Kathleen Brooks-Pazmany, *Research Assistant, National Air and Space Museum*
Gene Brown, *Holliston*
Fred Buonocore, *Down Inc.*
John C. Burton, *EAA Chief of Public Relations*
Ed Castle, *free-lance photographer*
Paul E. Ceruzzi, *Curator of Science and Technology, National Air and Space Museum*
Edward Chalkley, *Assistant Director for Operations, Preservation, Restoration, and Storage Division*
Ross Chapple, *free-lance photographer*
Rita Cipalla, *Chief, Public Affairs and Museum Services, National Air and Space Museum*
Anthony Collins, *Colotone, Inc.*
Calvin Cornelious, *Carver Photocomposition, Inc.*
Royce Culbertson, *Holliston*
R.E.G. Davies, *Curator, National Air and Space Museum*
Liz Early, *Colotone, Inc.*
Bill Fawcett, *Delta Mills Marketing Company*
Mary Henderson, *Curator of Art, National Air and Space Museum*
Bill Hezlep, *cartographer*
Peter L. Jakab, *Historian, National Air and Space Museum*
Carol James, *Smithsonian Docent*
Robert F. Jillson, *Holliston*
Pete Jurgaitis, *Lehigh Press Lithographers*
John G. King, *R. R. Donnelley & Sons Company*
Janie P. Koussis, *Watermark, Inc.*
George C. Larson, *Editor, Air & Space Magazine*
Susan Lawson, *Museum Technician, Art, National Air and Space Museum*
Russell E. Lee, *Curatorial Assistant, National Air and Space Museum*
Donald S. Lopez, *Deputy Director, National Air and Space Museum*
Mary Ellen McCaffrey, *Production Control Officer, Smithsonian Institution Photo Services*
Mark McCandlish, *artist*
Mary McElveen, *General Electric*
Ron Moore, *Carver Photocomposition, Inc.*
Lester Myers, *Museum Protection Officer, NASM*
Fred Nagel, *R. R. Donnelley & Sons Company*
A. Nailer, *Librarian, The Royal Aeronautical Society*
Annie Nelson, *Security Aide, NASM*
William Neyman, *Carver Photocomposition, Inc.*
Stefan Nicolaou, *Musée de l'Air*
Sam Nutwell, *Stanford Paper Company*
Barbara O'Malley, *Smithsonian Docent*
Jeff Ogden, *Carver Photocomposition, Inc.*
Norberto Parris, *Museum Protection Officer, NASM*
Roberto Pulos, *Mobile Equipment Operator Leader, National Air and Space Museum*
Walter Roderick, *Chief, Production, Operations, Preservations, Restoration and Storage Division, National Air and Space Museum*
Karl S. Schneide, *Curatorial Assistant, National Air and Space Museum*
Caroline Sheen, *free-lance photographer*
Pete Suthard, *Chief, Information Management Division, National Air and Space Museum*
James Trimble, *National Archives*
Steve True, *Lindenmeyr Paper Corporation*
Robert van der Linden, *Assistant Curator, National Air and Space Museum*
Estelle Washington, *Mobile Equipment Operator Leader, National Air and Space Museum*
Ed Watters, *The Lehigh Press, Inc.*
Don Wheeler, *Watermark, Inc.*
Ivory J. Williams, *Museum Protection Officer, NASM*
Lawrence E. Wilson, *Technical Information Specialist, National Air and Space Museum*
Ad Graphics
Federal Express
Type Studio
Virginia Bader Fine Arts Limited

Picture Credits

Legend: B Bottom; C Center; L Left; R Right; T Top.

The following abbreviations are used to identify Smithsonian Institution museums and other collections.

SI Smithsonian Institution; NASM National Air & Space Museum; ADIII Alexis Doster III Collection; EAA Experimental Aircraft Association; IWM Trustees of the Imperial War Museum; LC Library of Congress; ILN Illustrated London News Picture Library; NA National Archives.

Front Matter p. 1 NASM/SI, photo by Ross Chapple; 2–3 NASM/SI; 4–5 *Victory Salute*, Robert Taylor, reproduced by kind permission of Mr. Nick Maggos & The Military Gallery of Great Britain, Alexandria, VA; 6–7 George Hall/Woodfin Camp & Assoc.; 8 NASM/SI; 8–9 Fred J. Maroon; 9 NASM/SI; 10–11 Ray Miller; 12–13 *Afterglow*, Naima Rauam, NASM/SI, gift of Wendy Stehling; 15 Caroline Sheen.

The Once and Future AIRSHOW! pp. 16 Seaver Center, L.A. County Museum of Natural History; 17 Jim Koepnick/EAA; 18T Joe B. Goodwin; 18B Kenneth Garrett; 19T Frances C. Rowsell; 19C Kenneth Garrett; 19B Joe B. Goodwin; 20–21 Jim Koepnick/EAA; 22 Budd Davisson; 23 Caroline Sheen; 24 Chris Sorensen; 24–25 Ron Page/EAA; 25 Budd Davisson; 26T Kenneth Garrett; 26B Joe B. Goodwin; 26–27 Bill Johnston/EAA; 27 Joe B. Goodwin.

Part 1 From Dream to Reality pp. 28–29 LC; 31 NASM/SI.

False Starts and Triumphs pp. 32 NASM/SI, photo by Ed Castle; 33 NASM/SI; 34–37 art by Paul Takacs, photo by Ed Castle; 38,39 NASM/SI; 40 The Bettmann Archive; 41T courtesy Musée de l'Air; 41B NASM/SI; 42,43 NASM/SI; 44 LC; 45T NASM; 45B LC; 46 Dept. of Archives & Special Collections, Wright State University, Ohio; 47,48 LC; 49, 50–51 NASM/SI; 52 Charles H. Phillips; 53 LC.

America's Forfeited Legacy pp. 54 courtesy Musée de l'Air; 55 NASM/SI; 56–57 LC; 58,59 NASM/SI; 60T The Bettmann Archive; 60–61,61R NASM/SI; 63T LC; 63B The Bettmann Archive; 64 NASM/SI; 64–65 L'Illustration/SYGMA; 65 NASM/SI; 67 Air Portraits Colour Library; 68 NASM/SI Libraries; 69T *Harriet Quimby, 1911*, Flohri, NASM/SI; 69B,70T NASM/SI; 70B courtesy Museo Aeronautico Caproni di Taliedo, NASM/SI; 71 NASM/SI, photo by Ross Chapple.

Part 2 Two Wars and Between pp. 72–73 The Bettmann Archive; 74–75 NASM/SI; 75 *NC-4*, Ted Wilbur, 1976, NASM/SI, gift of Stuart M. Speiser; 76–77 courtesy The Boeing Company Archives; 77T NASM/SI; 77B courtesy Lockheed-California Co.

WWI: The Accelerator pp. 78–79 IWM; 79 NASM/SI, photo by Ed Castle; 80,81 IWM; 82 Musée de l'Air; 83 ILN; 84 Musée de l'Air; 84–85 *Aerial Resupply*, Charles L. Lock, 1980, U.S. Marine Corps Art Museum; 86–87 *Aeroplane Fight Over the Verdun Front*, Henri Farré, courtesy USAF Art Collection; 88 The Bettmann Archive; 89 From: *Eye Deep in Hell* by John Ellis. Croon Helm Ltd, 1976, London, photo by Ed Castle; 90T NA; 90B NASM/SI; 91T NASM/SI, photo by Ed Castle; 91CL,CR,BR NA; 91BL Musée de l'Air; 92 NA; 93T L'Illustration/ SYGMA; 93B NASM/SI; 94 Süddeutscher Verlag Bilderdienst; 95T NA; 95B, 96IWM; 97 NASM/SI; 98 art by Greg High, photo by Ed Castle; 99 NASM/SI, photo by Ed Castle; 100 NA; 101 The Bettmann Archive; 102 NA; 103T LC; 103B NA; 104T NA; 104B ADIII, photo by Ed Castle; 104–105,107 NA; 108,109 NASM/SI; 110–111 NASM/SI, photo by Ross Chapple; 112 IWM; 113 *Allies Day, May 1917*, Childe Hassam; National Gallery of Art, Washington, DC, gift of Ethelyn McKinney in memory of her brother, Glenn Ford McKinney.

Boom and Bust pp. 114–115 *Threatening Weather, But the Mail Must Go Through 1924*, Wilma Wethington, 1983, NASM/SI, gift of the artist; 115,116 NASM/SI; 117T The Bettmann Archive; 117B NASM/SI, photo by Ed Castle; 118 The Bettmann Archive; 119L courtesy Association of Flight Attendants; 119R Fred Winkowski; 120,121 NASM/SI; 122 The Bettmann Archive; 123,124 NASM/SI; 124–125 NASM/SI, photo by Ross Chapple; 126 NASM/SI; 127 NASM/SI, photo by Ross Chapple; 128T Bella Landauer Sheet Music Collection, NASM/SI Libraries; 128B NASM/SI, photo by Ed Castle; 129 L'Illustration/SYGMA; 130–131 detail from *Speed of Transportation*, Herman Sachs, 1929, photo by Randy Juster; 131T NASM/SI.

The Golden Age pp. 132–133 NASM/SI, photo by Ross Chapple; 133 Ed Castle; 134,135 NASM/SI, photos by Ed Castle; 136–137 *Picnic with Travel Air*, David Zlotkey, NASM/SI; 138 *Preparing for Atlantic Survey, 1933*, John Paul Jones, NASM/SI, gift of the artist; 139L NASM/SI; 139R Bella Landauer Sheet Music Collection, NASM/SI Libraries; 140,140–141 courtesy Sikorsky Aircraft; 142 *Portrait of Igor Sikorsky*, Victor Olson, 1974, NASM/SI, gift of the artist; 143L courtesy Sikorsky Aircraft; 143R, 144 NASM/SI; 145 NASM/SI, photos by Ed Castle; 146–147 *Hughes Racer*, Bruce Burk, NASM/SI, gift of the artist; 148L Freelance Photographers Guild; 148R NASM/SI; 149TL,B NASM/SI; 149TR © 1934 Time Inc. All rights reserved. Reprinted by permission from TIME; 150T courtesy The Royal Aeronautical Society; 150B *Hawker Harts Over the Himalayas*, Frank Wootton, reprinted by kind permission of the artist; 151T NASM/SI; 151B courtesy The Royal Aeronautical Society; 152 art by Greg High, photo by Ed Castle; 153 NASM/SI; 154–155 *Curtiss SOC-1, Scouting 3, USS Mississippi*, R.G. Smith, NASM/SI, gift of MPB Corp.; 155 Steve Cox, NASM/SI, gift of the artist; 156 NASM/SI; 156–157 NASM/SI, photo by Ross Chapple.

WWII: Reaping the Whirlwind pp. 158 *Hans-Ulrich Rudel at Work*, Keith Ferris; 159 Polska Agencja Interpress; 160 NASM/SI; 161 NA; 162–163 *Battle of Britain*, Frank Wootton, 1983, reproduced by kind permission of the artist; 163T ADIII, photo by Ed Castle; 164–165 Image in Industry Ltd., photo by Arthur Gibson; 166 Tom Lea, offical U.S. Army portrait; 167 *Into the Teeth of the Tiger*, William S. Phillips, © 1984, reprinted by permission of Greenwich Workshop, Inc. CT, on loan to NASM/SI by Donald S. Lopez; 168T *Tora! Tora! Tora!*, Robert McCall, reproduced courtesy of the artist, on loan to NASM/SI, photo by Ed Castle; 168B Robert Hunt Library; 169 LC; 170 *Doolittle Raid*, Keith Ferris, 1974. On permanent loan to NASM/SI; 172 Peter Stackpole, LIFE Magazine © 1942 Time Inc.; 172–173 *Douglas SBD-3*, R.G. Smith, NASM/SI, gift of MPB Corp.; 174 Franklin D. Roosevelt Library; 174–175 courtesy General Dynamics; 175 George Strock, LIFE Magazine © 1940 Time Inc.; 176 courtesy Lockheed-California Co.; 177L Terry Gwynn-Jones; 177R *Adm. Isoroku Yamamoto, I.J.W.*, Shugaku Homma, 1943, courtesy Naval Historical Center; 178 *Republic P-47D Thunderbolt*, Keith Ferris, 1974; 179T NA; 179B NASM/SI, USAF collection; 180–181 NASM/SI, photo by Ross Chapple; 181 Margaret Bourke-White, LIFE Magazine © 1943 Time Inc.; 182L ILN; 182R ADIII, photo by Ed Castle; 183 *Royal Air Force 617 Squadron Raid on the Möhne Dam, May 16, 1943*, Frank Wootton, on loan to NASM/SI, reproduced by kind permission of the artist; 184,185 ADIII, photos by Ed Castle; 186 NASM/SI Libraries; 187 ADIII, photo by Ed Castle; 188–189 Bill Crump/Confederate Air Force; 191 *Götterdämmerung* (Twilight of the Gods), Harold Schmidt, NASM/SI, photo by Ed Castle; 192 NASM/SI, photo by Ed Castle. *Gatefold*: Right, *Fortress Under Fire*, Keith Ferris, NASM/SI; Center, From: *Atlante Enciclopedico degli Aerei Civili del Mondo da Leonardo a oggi* © 1981 by Arnoldo Mondadori Editore S.p.A., Milano & *Atlante Enciclopedico degli Aerei Militari del Mondo dal 1914 a oggi* © 1980 by Arnoldo Mondadori Editore S.p.A., Milano & NASM/SI Libraries. Left, NASM/SI & private collections, photo by Ed Castle. 193 *Unsung*, James Dietz, courtesy Office of the Secretary of the Navy, on loan to NASM/SI; 194–195 *Lucky Strike*, William S. Phillips, NASM/SI, gift of Warren & Phylis Bush; 195 NASM/SI, USAF collection; 196 US Naval Academy Museum, Annapolis; 197 Carl Mydans, LIFE Magazine © 1945 Time Inc.

Part 3 Jets, Rockets and Realism pp. 198–199 NASM/SI, photo by Ross Chapple; 201 Louis Bencze.

The Jet Age pp. 202–203 NASM/SI, photo by Ross Chapple; 203 NASM/SI; 204 *Beech 18*, Keith Ferris, 1980; 204–205 Herman J. Kokojan/Black Star; 207 Popular Mechanics Magazine; 208 NASM/SI; 209 courtesy Lockheed-California Co.; 210, 211, 212, 213, 214, 214–215 NASM/SI; 215 Michael Montfort/Woodfin Camp & Assoc.; 216L courtesy Rockwell International; 216R art by Greg High, photo by Ed Castle; 217 *MiG Might*, Mark McCandlish, 1986, donated to San Diego Aerospace Museum, photo by Ed Castle; 218T de Havilland/Air Pilots Assoc.; 218B Keystone Press Agency; 219 de Havilland, Air Pilots Assoc.; 220, 220–221,222 NASM/SI; 223 courtesy General Dynamics; 224–225 Jim Pozarik/Gamma-Liaison Agency; 225 UPI/Bettmann Archive Newsphoto.

Bigger, Better, Faster pp. 226–227 Erik Simonsen; 227 *AV-8 Harrier*, Joe Plummer, 1984; 228–229 Chourgnoz/SYGMA; 230 map by Bill L. Hezlep; 231 Image in Industry, Ltd., photo by Arthur Gibson; 232T courtesy Sikorsky Aircraft; 232B courtesy Lockheed-California Co.; 233 Jose Fernandez/Woodfin Camp & Assoc.; 234 art by Greg High, photo by Ed Castle; 235 *Thunderchief F-105*, Michael Turner; 236 courtesy Bell Helicopter/Textron; 237T *Heading for Trouble*, William S. Phillips, © 1985. Reproduced by permission of Greenwich Workshop Inc., CT; 237B Robert Ellison/Black Star; 238 Fred J. Maroon; 238–239, 239R George Hall/Woodfin Camp & Assoc.; 240T *Bekaa Valley Gunfight*, Robert Taylor, reproduced by kind permission of The Military Gallery of Great Britain, Alexandria VA; 240B NASM/SI; 241T art by Greg High, photo by Ed Castle; 241B USAF/Walter J. Boyne; 242 art by Greg High, photo by Ed Castle; 243 Michael Melford/Wheeler Pictures; 244T Karen Kasmauski/Wheeler Pictures; 244B Roger Ressmeyer; 245 Co Rentmeester/Image Bank; 246 Michael Melford/Wheeler Pictures; 246–247 *Turning Home*, © Attila Hejja, 1981, Royal Saudi Air Force Art Collection; 247 Michael Melford/Wheeler Pictures; 248–249 Fred J. Maroon; 250T P. Kyriazis/SYGMA; 250B courtesy Grumman Corp.; 251 U.S. Dept. of Defense; 252–253 James Sugar/Black Star; 254T AP/Wide World Photos; 254B NASM/SI; 255T *The Lightship*, © Attila Hejja, 1984. NASA Fine Art Collection; 255B *2001 and Beyond*, © Attila Hejja, 1982, courtesy Loral Corp/USAF Art Collection; 256 Air Portraits Colour Library; 257T courtesy Sikorsky Aircraft; 257B courtesy G.E. Air Engines; 258–259 Michael Melford/Wheeler Pictures; 260 Paul Chesley/Photographers Aspen; 260–261 David O. Hill/Photo Researchers.

Flight to the Future pp. 262–263 J.P. Laffont/ SYGMA; 263 Farrell Grehan/Photo Researchers; 264 NASM/SI; 264–265 *Deperdussin and Gee Bee*, Guy Johnson, NASM/SI, The Stuart M. Speiser Photo-Realist Collection, gift of Stuart M. Speiser; 265T LC; 265B NASM/SI; 266–267 The Bettmann Archive; 267 NASM/SI; 268T Karen Petersen; 268B Mark Avino, NASM/SI; 269 NASM/SI, photo by Ross Chapple; 270T James Sugar/Black Star; 270B courtesy Grumman Corp; 271 courtesy British Aerospace, NASM/SI; 272 James Sugar/Black Star; 273 William J. Warren/West Light; 274 drawings by Paul MacCready; 275 William J. Warren/West Light; 276–277 *Smart Plane*, Attila Hejja, 1987, photo by Ed Castle; 278 *Solar sail*, Tom Hames/World Space Foundation; 278–279 Erik Simonsen.

Endpapers From *Aircraft Recognition*, Vol.1, October, 1942. NASM/SI, photo by Ed Castle.